Cross-Cultural Human-Computer Interaction and User Experience Design

A Semiotic Perspective

Cross-Cultural Human-Computer Interaction and User Experience Design

A Semiotic Perspective

Jan Brejcha

CRC Press
Taylor & Francis Group
Boca Raton London New York

CRC Press is an imprint of the
Taylor & Francis Group, an **informa** business

CRC Press
Taylor & Francis Group
6000 Broken Sound Parkway NW, Suite 300
Boca Raton, FL 33487-2742

First issued in hardback 2019

ISBN-13: 978-1-4987-0257-7 (hbk)

Library of Congress Cataloging-in-Publication Data

Brejcha, Jan.
    Cross-cultural human-computer interaction and user experience design : a semiotic perspective / Jan Brejcha.
       pages cm
    Includes bibliographical references and index.
    ISBN 978-1-4987-0257-7 (alk. paper)
    1. Human-computer interaction. 2. User interfaces (Computer systems)--Cross cultural studies. 3. Intercultural communication. 4. Communication in design. I. Title.

    QA76.9.H85B74 2015
    004.01'9--dc23
                                          2014032592

Visit the Taylor & Francis Web site at
http://www.taylorandfrancis.com

and the CRC Press Web site at
http://www.crcpress.com

Contents

PART I   Semiotics of Interaction

PART II   *Culture of Interaction*

PART III   *Appendices*

List of Figures

List of Tables

Preface

This book focuses on a semiotic approach in product, service, and system design, and emphasizes the semiotic and linguistic aspects in human-computer interaction (HCI) and user experience (UX). This work provides scientific information on the theoretical and practical areas of the interaction and communication design for research experts and industry practitioners from multidisciplinary backgrounds, including industrial designers, ethnographers, human-computer interaction researchers, human factors/usability engineers, UX/interaction designers, mobile product designers, and vehicle system designers. This book is organized into two parts, which focus on the following subjects:

I: Semiotics of Interaction
II: Culture of Interaction

Part I covers the theoretical background of HCI semiotics with emphasis on the interaction elements present in the user interface, the methodology to work with them to achieve useful insights both for design and evaluation, and the results we can obtain compared to traditional UX methods. Moreover, it focuses on the persuasive elements of HCI, and discusses various approaches for design and evaluation of interactive systems. Finally, it presents a case study showing the application of semiotic analysis to a complex set of user interfaces. Part II draws from the above theoretical approach, and applies it to a cross-cultural study presenting a semiotic method for design and evaluation in a different cultural background, and discusses the resulting insights, which are then structured as guidelines for HCI/UX design for Chinese users.

This book presents a novel approach to cross-cultural HCI/UX, covers the latest research in the field, and brings a set of tools and methods to benefit the design process by taking a semiotic perspective. I hope this book is helpful for the researchers and practitioners in developing more usable, useful, and appealing products, services, and systems.

Jan Brejcha
Prague, June 2014

Acknowledgments

The author acknowledges the assistance of Aaron Marcus and the AM+A library and document archive in providing information and examples for this text, and the assistance of the Sino-European Usability Center, namely of Prof. Zhengjie Liu, Gong Hong Yin, Han Li, Sun Wenxin, Ma Yin, Zhou Yongjie, Xiao Sheng, and Xiang Yong. Also, the author wishes to thank Dr. Denisa Kera and Dr. Jiří Bystřický for their advice, as well as the author's family for their patience and support in writing this book.

Author Biography

Jan Brejcha obtained his Ph.D. in HCI/UX at the Institute of Information Studies and Librarianship, Charles University in Prague, Czech Republic. During his studies he developed a cross-cultural semiotics and language-based design, analysis, and evaluation methodology, which is presented in this book. Jan leads HCI/UX-related lectures and workshops and engages as a consultant for interesting projects in the academic and corporate fields. His passion for human–machine interaction also drives him in new business development for a machinery trading company he co-owns. He capitalizes on his past studies and experience to design useful, usable, and appealing systems and services.

1 Introduction

"What pattern connects the crab to the lobster
and the orchid to the primrose
and all four of them to me? And me to you?"

Gregory Bateson, *Mind and Nature*

The present work seeks to describe implicit patterns based on natural language and culture in human-computer interaction (HCI), and user experience (UX). For our purposes we shall use the term HCI as concerned about the structure of the system, while the term UX shall be used for the activities related to the HCI design process, such as planning, researching, analyzing, designing, implementing, evaluating, documenting, training, and maintaining (Marcus, 2009). The user interface (UI) is the external effect (or representation) of both HCI, and UX.

Although extensive research has been done in the fields of HCI semiotics, and the cultural aspects of HCI, we seek to bring novel insights by informing these fields through the linguistic perspective, and to propose a set of design guidelines for the international HCI/UX practitioners. The goal to achieve this is twofold: First, to define the semiotic and linguistic system of HCI and create a UX methodology (Part I), and second, to apply the acquired methodology for cross-cultural comparison (Part II).

We consider semiotics as a foundation of interaction and communication design in HCI, because it is concerned with meaningful arrangement of UI elements across space and time.

However, the above method can work well only when we acknowledge the influence of our native language and culture on our thoughts and actions. The potential differences are further accented by globalization, because when using communication technology, we are faced more and more with UIs coming from rather different cultural backgrounds. In order to tackle the differences in a meaningful manner, there is a growing need to design UIs that are usable and well accepted in the target culture.

Cross-cultural testing of UIs is the most comprehensive way to meet this goal, but it is also the most financially demanding. Therefore, by defining a usable set of UI design guidelines for a target culture, designers could market their products with lower costs and with better acceptance.

We chose to work from the semiotic perspective, which helps us uncover the sense-making processes of the users. We used semiotic methods to build a common framework to gather and analyze cross-cultural data. And, we started our analysis by looking at the UI as an example of complex language.

To acquire the necessary preliminary insights about users from diverse cultural backgrounds, we carried a pilot study targeted at the habits, mental models, and UI preferences of Chinese and Czech users. The scientific rationale of choosing Czech and Chinese respondents for comparison was to search for meaningful differences between cultures, which could be turned into usable design guidelines for the UX community. The two groups were chosen, first, because one is a part of the Western culture, and the other is a part of the Eastern culture. Previous literature showed interesting differences related to HCI between these two groups. Second, the author was able to conduct the research in these two groups according to the method presented above, thanks to his direct access.

In our work we followed and expanded upon a body of previous research in the field of cross-cultural research. In our view, however, only limited work has been done in creating usable guidelines for cross-cultural UI design. We bring our insights from our cross-cultural research and propose a set of design guidelines.

The present book aims at analyzing HCI/UX following these main theses:

1. **The UI is a means of sharing and interpreting information between systems.** Our thoughts and actions are guided by intrinsic logic rules, supported by the system of language and culture. Language provides an architecture of the design space of HCI/UX. Linguistics and semiotics provides effective methods to solve problems in communication and interaction design. These methods help define the users in their culture, rather than as culture-independent agents. Moreover, each UI stands on a certain paradigm of use which is not always apparent. The HCI ideology defines what relations between users and objects can (or should) be made.

2. **Every sign in HCI is cultural and therefore informational.** The UI provides a lens for reading and writing cultural data. The user's native language and culture determines his/her mentality, rationality, and the discourse involved. By expressing in different systems of meaning (e.g., languages, UIs), we accent different objects and experiences, which results in different insights into the world we live in. When UIs take into account those differences, they can promote both usability and cultural diversity.

The theoretical backbone of our work can be illustrated by the following concepts and their relations:[1]

Data -> Facts -> Thought/design – **Picture of reality**/ideology <-> Form/ design – **Language**/culture <-> **Information** <-> Language/UI <-> **Interaction**.

Data are factual information (Merriam-Webster, 2013a); facts are the existence of states of affairs (Wittgenstein, 1922, p. 25). Out of the sensory data a person receives, he or she interprets the world to make sense of it ("The world is the totality of facts…" Ibid.). We are used to structure the interpreted data, or facts, in relations that make sense to us. As Wittgenstein reminds us, "The logical picture of the facts is the thought" (Ibid., p. 30). Logic comes from the Greek *logos*, which stands for speech, word, and reason; it is the "controlling principle of the universe"

---

[1] The relations are inspired by Hjelmslev (1961), and we use the following symbols: "->" for determination, "–" for constellation, "<->" for interdependency.

(MerriamWebster, 2013b). Although different systems and entities, such as humans and computers, interpret data differently, they are able to share certain logical structures of objects and processes. Thanks to this unifying principle, our natural language can create various programming languages, which strictly follow a set of logical rules, allowing them to be readily interpreted and executed by the computer. The execution of commands given to the computer is analogous to our utterance of commands, as is the case of speech acts. Speech acts were coined by Austin (1962) for utterances performing actions in the world (e.g., when promising or ordering something).

Thought has therefore a strong relation with logic. "The thought is the significant proposition" (Wittgenstein, 1922, p. 38), that is, a proposition with a sense. Wittgenstein suggests language is "[t]he totality of propositions" (Ibid.). What are the aspects of language we shall work with? According to Searle, language possesses the following three aspects: qualitativeness, subjectivity, and unity. Qualitativeness is a "character of conscious thoughts, or qualia" (Searle, 2002, p. 40). Subjectivity is a "first-person ontology of subjective conscious states, because they exist only when they are experienced by some human or animal agent" (Ibid., p. 41). For unity, "all conscious experiences at any given point in an agent's life come as part of one unified conscious field [...] A conscious state is by definition unified, and the unity will follow from the subjectivity and qualitativeness" (Ibid.). Natural language is characterized by discreteness, compositionality, and generativity (Searle, 2009, pp. 63–64). A well-defined UI language should also have these features, in order to work properly. For the linguistic and semiotic perspectives of HCI/UX design see Part I, "Semiotics of Interaction."

Thought thus arranges facts as pictures in the processing of reasoning. This means the thought connects within itself a logical system of language together with its depiction, or model, because "[t]he picture is a model of reality" (Wittgenstein, 1922, p. 28). Thought as a reasoning and logical process provides thus a link between systems that are mutually untranslatable, but which interact with one another. We can notice the link in the above example with natural language and the programming language, which stands behind the UI. Thought connects also the textual and pictorial expression with design. The UI "is always an effect. It is always a process or a translation" between significant expressions (Galloway, 2008, p. 939). The UI stands on a system of ideas and beliefs, an HCI ideology.

Thought joins ideology with action. The nexus of both of them finds its place in the UI. The UI as an artifact has a strong analogy with architecture, as it is presented by Somol and Whiting (2002). According to them, architecture can be regarded as an index, or "mediator [which combines] materialism with signification" (Ibid., p. 74). "Architecture is both substance and act. The sign is a record of an intervention—an event and an act [...]" (Ibid.). The UI as a mediator presupposes a user to work with it, thus enacting the ideology behind it. In the context of theater this is known as a parallax. "[P]arallax is the theatrical effect of a peripatetic view of an object. It takes into account how the context and the viewer 'complete' the work of art" (Ibid., p. 76). While parallax is purely optical, the Doppler effect is more general. "[The Doppler effect is] an atmospheric interaction. It foregrounds the belief that both the subject and the object carry and exchange information and energy. In short, a user might be more attuned to certain aspects of a building than others" (Ibid.). And that is why architecture can provide important implications for the design space of HCI/UX. The HCI ideologies present in the UI are discussed in Section 2.3, "Ideology, persuasion."

Thought, or consciousness, and all mental phenomena are "higher level features of the brain," that are "caused by lower level neurobiological processes in the brain" (Searle, 2002, p. 18). As such, consciousness processes sensory data from different sources. Searle continues, that consciousness and intentionality build language (Searle, 2009, p. 64). In the context of the present work, we conceive intention as design, and intentionality as a designing force. Therefore, for the purposes of our work, we understand the linguistic elements present in the UI (or, the UI language) as a product of a thinking process of a designer. Thought can produce designs, or we can identify it with design itself. As etymology reminds us, design comes from the Latin *designare,* to mark out; composed from *de-* + *signare* to mark (Merriam-Webster, 2013c). As a verb it means to conceive or execute a plan or a scheme, and as a noun it stands for a project, sketch, or pattern. By marking a material we leave a sign in it. Therefore, we inform it. Design is thus the nexus between the immaterial (intention) and the material (matter) that informs (from the Latin *informare,* see MerriamWebster, 2013f). By giving a form or a character to something, we communicate information. A similar argument is put forward by Flusser (1999). Following the above definitions, we see design as closely related to semiotics (through signs and intentionality) and to information science (through information). In the context of UX, design finds its application in industrial design, interaction design, and communication design. Industrial design is related to material (tangible) artifacts, whose operation is guided by interaction design. The latter is supported by communication design through immaterial cues, which inform the users on the operation results. In this work we shall focus on the dynamic and immaterial applications of design, i.e., interaction and communication design.

Language constitutes a frame for thought and experience by creating conventions. Conventions are arbitrary, but once they are settled they give the participants a right to specific expectations. They are normative (Searle, 2009, p. 87). Conventions lead to institutions. "Language is essentially constitutive of institutional reality" (Searle, 1995, p. 59). Language completely interpenetrates with experience (Sapir, 1949, p. 11). Therefore, when a designer wants to create a certain UX, he or she should constitute it primarily through the system of UI language. And, by that the designer acts upon a user's natural language and thought. Because language creates cultural conventions, culture stands therefore on natural language. Language is used for cultural accumulation and historical transmission: proverbs, medicine formulae, standardized prayers, folk tales, standardized speeches, song texts, genealogies, etc. (Ibid., p. 17). "Every complex of a culture and a lge [language] carries with it an implicit metaphysics; a model of the universe, composed of notions and assumptions organized into a harmonious system which is valid for framing statements about what goes on in the world as the carriers of the culture see it" (Whorf, 2012, p. 361). This model of the universe sets the basic ideological frame for our interaction, both in the social context, as well as the computer context. Social interaction is crucial for the experience with the environment, because the interaction is processed in natural language. According to Sapir, in a society "even the simplest environmental influence is either supported or transformed by social forces. Hence any attempt to consider even the simplest element of culture as due solely to the influence of environment must be termed misleading" (Sapir, 1949, p. 89). Therefore, every UI with its designers and users has to be scrutinized, not only from the perspective of HCI, but also from the perspective of culture of the HCI participants. The cultural differences in interaction are discussed in Part II, "Culture of Interaction."

Part I

Semiotics of Interaction

Parts of this section are used with the kind permission from Springer Science+Business Media: Towards a UI Alphabet In: HCI International 2013 Conference Proceedings by Springer in the Lecture Notes in Computer Science (LNCS) series. The original publication is available at http://www.springerlink.com.

2 Semiotic Foundations of HCI and UX Design

As we have described in our thinking and acting, natural language plays a central part. This language defines a structure even before we form something and can be regarded as the architecture of design. Our consciousness is the result of language informing design. Grammar allows for many combinations of objects and actions, but ideology establishes the privileged connection of the two of what is correct and possible. In order to set forth the privileged connection, ideology employs different forms of persuasion. To grasp the expression of these structures in HCI, we chose the perspective of linguistics and semiotics.

By semiotics, we mean a theory of signs. We combine the Anglo-American semiotics (semeiotics) perspective with the French semiology ("sémiologie") approach (Barthes, 1977). According to Peirce, a sign is "something that stands for someone or something in some respect or capacity" (Peirce in Buchler, 1955, p. 99). Four dimensions form the sign: lexical (Eco, 1979), syntactics, semantics, and pragmatics (Morris, 1970).

The lexical dimension (Eco, 1979) focuses on the generation of elements and their syntax. The lexical dimension is bound to the physical constraints (see Norman, 1999) of the UI, such as display size, resolution, and material used. The lexical dimension is related to consumption (Marcus, 2009).

Syntactics is "the study of the syntactical relations of signs to one another in abstraction from the relations of signs to objects or to interepreters. . . " (Morris, 1970, p. 13). In this dimension we deal with the grammar constituting relations between the perceivable elements or sign vehicles.

Semantics, on the other hand, "deals with the relation of signs to their designata and so to the objects which they may or do denote" (Morris, 1970, p. 21). This dimension is devoted to the relation between vehiculae and the object, content, action, or "meaning" the UI represents and enables. This dimension is connected to mental models (in the sense of which functions the system allows) and affordances (Norman, 2002; Gibson, 1977).

Pragmatics "deals with the biotic aspects of semiosis, that is, with all the psychological, biological, and sociological phenomena which occur in the functioning of signs" (Morris, 1970, p. 30). This most complex dimension focuses on how we use or interpret the vehicula-object relation, that is, what is the sign's purpose? The pragmatic dimension governs how signs are used, or understood in their conventional and symbolic form.

We consider semiotics as a foundation of interaction and communication design in HCI, because it is concerned with meaningful arrangement of UI elements across space and time. The semiotics perspective in the context of HCI is increasingly popular in presenting a different approach. The classical linguistic and semiotic foundations of HCI were previously set down by, e.g., Andersen, 1997, 2001; Brandt, 1993; Nadin,

1988; de Souza, 2005), and to other disciplines as well, for example, typography (Ehses, 1976). In the following chapters we take their contribution to build a set of semiotic heuristics that we use to evaluate a complex UI example. We present a semiotic evaluation method and report the results of our in-depth investigation. Our semio-linguistic analysis takes into account also the pragmatics of HCI, manifested through rhetorical tropes to persuade the user.

2.1   SIGNS, DATA, INFORMATION

The signs presented in the context of interaction and communication design are computed and manipulated as data. This data, or a *datum*, is a lack of uniformity (Floridi, 2010, p. 23). The lack of uniformity stands at the very outset of sign creation. This process is known as coding (see, e.g., Flusser, 2002, p. 53). In HCI, there are two types of coding involved: "1) the creation, change, and interpretation of the shape of signs; and 2) the storage, linkage, and retrieval of signs into, in, and from memory" (Pearson and Slamecka, 1977, p. 40). This communication process lies at the heart of information theory that "studies the problems of signal transmission, reception, and processing" (Merriam-Webster, 2013e). "Information in its everyday sense is a qualitative concept associated with meaning and news. However, in the theory of information, it is a technical term, which describes only quantifiable aspects of messages. Information theory and semiotics have goals of similar analytic universality: Both study messages of any kind, but because of its strictly quantitative approach, information theory is much more restrictive in its scope" (Nöth, 1995, p. 34). This confirms also Moles (1966) by writing about quantifying information: "the measure of information must be based on originality and not on signification" (Ibid., p. 22). For a message to be meaningful, it must have a high level of sign redundancy. Therefore, the more meaningful a message is, the less original it is. The measure of information included is also the object of one of Grice's (1975) maxims, namely that of the category of quantity:

1. Make your contribution as informative as is required (for the current purposes of the exchange).
2. Do not make your contribution more informative than is required.

(Grice, 1975, p. 45)

The maxims of quantity relate to the information presented in the UI. It is the dialog messages, choice of menu items, etc., which should be kept minimal. Further, the category of quality deals with the truthfulness of information:

1. Do not say what you believe to be false.
2. Do not say that for which you lack adequate evidence.

(Grice, 1975, p. 46)

Bringing the quality maxims more in the HCI/UX context, we realize the systems should present not only valid information, but also information relevant to the interaction context. For our purposes the category of manner is also important, which aims at presenting clear, well-structured information:

1. Avoid obscurity of expression.
2. Avoid ambiguity.
3. Be brief (avoid unnecessary prolixity).
4. Be orderly.

(Grice, 1975, p. 46)

Information theory is thus concerned with the lexical and syntactic dimension of semiotics. "The role that signs play in information processes (that is, in semiotic interactions) is determined by the properties of the sign; in turn, sign properties are determined by the kind of sign and its structure" (Pearson and Slamecka, 1977, p. 5). There is a close resemblance with the communication model first described by Shannon (Shannon, 1948, p. 381). If we focus solely on the lexical and syntactic dimensions of the model, we get an information source as the first communication component, followed by an encoder, and the physical medium of a communication channel. In the semiotic interaction "we first generate the semiotic context of a sign for communication; next, we add a shape to the sign and its context; and finally, we embody the sign in some physical medium so that the communication can actually be carried out" (Pearson and Slamecka, 1977, p. 26).

In HCI/UX design we are choosing signs that our users (or audience) would be able to perceive well, because "signs are the vehicle of perception, and the denotata of signs are the objects of perception. Perception as a semiotic, or information, process is similar to communication" (Pearson and Slamecka, 1977, p. 27). By perceiving, we already participate in communication and interaction. *"To perceive is to select; to apprehend the world is to learn the rules of perceptual selection"* (Moles, 1966, p. 60).

This perspective is shared with information science, which "deals with the processes of storing and transferring information" (Merriam-Webster, 2013d). In this context the basic sign in HCI is information. Here, we are entering into the semantic dimension of semiotics, because information, in general terms, is the unity of data and meaning (Floridi, 2010, p. 20). It is important to note, however, that the data "constituting information can be meaningful independently of an informee" (Ibid., p. 22). Raber and Budd (2003), for example, views semiotics and information science as concerned with representation and the production of culture. They conclude, that "the relationship between representation and what is being represented, are at the heart of information science" (Ibid., p. 225). The cultural background of the communicators determine which information can be transmitted: "(a) *Semantic* information, having a universal logic, structured, articulable, translatable into a foreign language, serves in the behaviorist conception to prepare *actions*. (b) Instead of to a universal repertoire, *esthetic* information, which is untranslatable, refers to the repertoire of knowledge common to the particular transmitter and particular receptor" (Moles, 1966, p. 129). Moreover, "[e]sthetic information is specific to the *channel* which transmits it; it is profoundly changed by being transferred from one channel to another" (Ibid., p. 131), for example, reading a printed book conveys different information than reading it online, or by listening to an audiobook.

The representational and cultural production role of semiotics and information science reduces uncertainty in communication and interaction, or entropy. Entropy is "the measure of the amount of information conveyed by a message from a source. The more we know about what message the source will produce, the less uncertainty, or entropy, and the less the information" (Pierce, 1980, p. 23). Entropy can be reduced also by excluding, what is meaningless to the recipient. As Luhmann (1995) points out, humans reduce complexity through meaning. "Language is a medium distinguished by the use of signs[. . .]. This leads to problems of complexity that are solved by rules for the use of signs, by reducing complexity, and by settling into a bounded combinatory capability" (Ibid., p. 160), that is, complexity in HCI can be reduced by implementing a grammar for visible and interaction elements (for details see Section 2.4, "UI languages"). Such grammar builds user's expectations by structuring the meaning (semantic dimension) of signs. As Luhmann further points out, "every specific meaning qualifies itself by suggesting specific possibilities of connection and making others improbable [. . .]. Meaning is consequently—in form, not in content—the rendering of complexity, that [. . .] permits access at a given point but that simultaneously identifies every such access as a selection" (Ibid., p. 61). Moreover, "[n]ot all systems process complexity and self-reference in the form of meaning; but for those that do, it is the *only* possibility. Meaning becomes for them the form of the world and consequently overlaps the difference between system and environment" (Ibid.). In HCI, it is analogous to a perception of a flow in immersive interaction.

2.2   ACTORS, AUDIENCES, PARADIGMS

In regard to analyzing the UI in terms of language we use other related disciplines, namely linguistics. We build upon the argument, that "linguistics is a part of semiotics," as stated by Cassirer (1945, p. 115; cited from Nöth, 1995, p. 229). Our semiotic method is based on the assumption that HCI concerns different actors (users, systems, designers) in a setting, or paradigm. Thus, the interaction can be seen as a form of discourse or conversation.

The process of HCI/UX design could be described as a form of storytelling, which makes us work mostly within the pragmatic dimension of semiotics. Here, we work with at least two actors. An actor can be the user or the system (computing agents); it can also be the designer of the system. There can be multiple users interacting with a system, as well as multiple systems interacting with the user. There can be users interacting among themselves directly or through the system. In this way we end up in a complex situation, which requires our attention on a technical, psychological, and social level of semiotics. Therefore it is here, where we discuss the work context, user's needs, and requirements. From the designer's perspective, the system he or she creates is directed towards the audience of the (potential) users. In this way we focus on the different types of users, and how to address them in the best possible (rhetorical) manner. This manner ranges from types of access to the system (or types of distribution), to the use of UI language, which we are going to define in Section 2.4, "UI languages." Here we intend to discuss market segments and personas. The created system takes the form of a paradigm in our context.

The paradigm sets the basic framework of interaction and communication by putting different types of constraints on the system, from the physical constraints of a specific device, semantic constraints of a software platform, to the constraints of the cultural background of deployment (see also Norman (2002)). The paradigm privileges, or hinders, different types of communication and interaction by the choice of discourses. We can define discourse as "a method for distributing information to prevent their entropy by nature" (Flusser, 2002, p. 15). Flusser describes various types of discourses in human history, ranging from theatrical discourse, hierarchical or pyramid discourse, scientific "tree" discourse, amphitheatrical discourse, democratic circular discourses, and network dialogs. "A dialog is a method for synthesizing a given information into new one" (Ibid., p. 22). In his view, HCI would be a combination of scientifically and technologically designed amphitheatres supported by networks of users (crowds), which change the previous way of communication (or our mind *program*), and shape a different societal structure (Ibid., p. 36).

In order to work within these paradigm shifts we should focus on the communication between users and the system, the ongoing narration. In HCI, a model of interaction can be exploited from narrative analysis. As Hébert (2006) put it, "The actantial model is a device that can theoretically be used to analyse any real or thematized action [. . .]." Hébert goes on to present the basic elements of the actantial model:

> The actantial model, developed by Greimas, allows us to break an action down into six facets, or actants: (1) The subject (for example, the Prince) is what wants or does not want to be joined to (2) an object (the rescued Princess, for example). (3) The sender (for example, the King) is what instigates the action, while the (4) receiver (for example, the King, the Princess, the Prince) is what benefits from it. Lastly, (5) a helper (for example, the magic sword, the horse, the Prince's courage) helps to accomplish the action, while (6) an opponent (the witch, the dragon, the Prince's fatigue or a suspicion of terror) hinders it.

While the actantial model could be fully used for example in game design, for other tasks we should work with a simplified structure. Brandt, for instance, developed (together with the notion of a semiotic interaction flow) a revised actantial model with the following modifications:

> (1) The subject controls the running of the program not only by starting and following it, but also by *feeding* it with input; the processing flow is sensitive to the input in quite a different way than, say, a water flow is senstive to modifications of its velocity: the "chemics" of the flow itself is modified by the input it receives [. . .] (it memorizes).
> (2) The "chemical" transformations of introduce input can be stopped by a bifurcation in the flow, until the user has been "asked" which way to proceed (it gives options).
> (3) But first of all, it "takes" informational *data*, by an inherent tendency to constipation and coagulation which breaks out intermittently and can only be cured by this specific remedy (it makes requests).

> (Brandt, 1993, pp. 134–5)

Brandt concludes, that "[t]he consequence of this semiotization of the flow (by the three features mentioned: memory, options, and requests) is a thorough revision of the actantial model. . . " (Brandt, 1993, p. 137).

So far we described a situation in which the user interacts (only) with the computer. How could this perspective change, when we start to consider also the designer of the UI? From the semiotic perspective it is important, that the user-computer interaction is a form of communication with the designer through the UI. The designer sends a "one-shot message" (de Souza, 2005, p. 84) through the UI to the user stating his or her intention and his or her understanding of the user's needs and wishes. The user then must decode this message in order to use the UI properly.

Because the designer is not present at the time of interaction, the UI plays the role of the "designer's deputy" (de Souza, 2005, p. 90). In the UI there are different texts and speech acts that the designer actually puts in and tells the user what to do, for example, in the form of documentation, help, tips, splash screens, intros, etc., which comes with the UI. The designer's deputy must be able to communicate the designer's intention (present in the one-shot message) of how the UI is to be used. In this case the metacommunication takes the form of narration that augments what the screens say to the user by themselves. These narrative parts can be structured as speech acts (Austin, 1962; Searle, 1969) and should follow the related characteristics (see Grice's maxims, 1975). As Winograd (1987–88) points out, "[s]peech acts are not unrelated events but participate in larger conversation structures [...]. An important example is the simple 'conversation for action,' in which one party (A) makes a request to another (B). The request is interpreted by each party as having certain conditions of satisfaction, which characterize a future course of actions by B" (Ibid., pp. 7–8). "The emphasis here is on language as an activity, not as the transmission of information or as the expression of thought" (Ibid., p. 10). Speech acts are a suitable tool for us, because they can take different forms. As Searle suggests, even just an action can be a speech act (Searle, 2009, p. 89). By using speech acts, we have the capacity to create a reality just by representing it. The conversation for action perspective is related to narration in Section 3.1.3, "Narration," and its implications for images are discussed in Section 2.4, "UI languages."

In this chapter we presented our main perspective that the designer generally communicates to the user through the UI at the moment of execution of the interaction. During the design process there is a two-way communication with the user, however, when the designer gathers facts and forms intuitions about the user's needs or wants. This notion is similar to the author and reader model, where the author conducts the reader through the text (Eco, 1979), and mental models of user, designer, and the system image (Norman, 2002). In the following chapter we shall discuss what forms this communication can acquire.

2.3   IDEOLOGY, PERSUASION

The UI of interactive systems is the meeting point of people with interactive communication technology (ICT). As a human product, it forms a part of culture that determines us, often without our full awareness. The UI is constructed according to a

Parts of this section are used with the kind permission from Springer Science+Business Media: Ideologies in HCI: A Semiotic Perspective In: HCI International 2014 Conference Proceedings by Springer in the Lecture Notes in Computer Science (LNCS) series. The original publication is available at http://www.springerlink.com.

set of values of the designer and other stakeholders in the production process. Their values and goals are implicitly encoded in the UI and the documentation, and can be in conflict with the values of the user. This means the UI directs the user interaction often according to the intent of the designer. This is when both the intentional and unintentional manipulation of the user starts because he or she is presented with choices, or even goals, that are incompatible for his or her intent. For the purpose of unmasking and decoding the inner workings of the UI, we have chosen a semiotics approach, with the emphasis on pragmatics, as defined above.

Each and every UI is a result of diverse influences during the design process. Stakeholders in the process have their own goals and expectations that they try to put into the final product. For example, the sales and marketing department could have strategic aims of short time-to-market, easy adoption of the product from the users and gimmicks to strengthen the brand and the product family. The programmers, on the other hand, might want to incorporate an advanced and clever technology, while the designers might want to create a simple and good-looking UI. All of these often conflicting values can have an input into the final product at the cost of the final user who expects the product to satisfy his or her needs and help achieve his or her goals. Often, such expectations fall short and the user is forced to become a "detective" trying to guess the motive of the designer in order to understand how to use the product in a sensible way (de Souza, 2005). In this light, the user should be aware as much as possible of the techniques used during the development process, as well as the prevailing HCI ideologies driving the UI production. Some even argue for a philosophy of technology:

> ... when HCI was primarily concerned with issues of usability, the question of what was a "good design" could be defined clearly; the time it took to complete a task, the error rate, or the learning curve. (...) To understand what makes a "good user experience," HCI will need a philosophy of technology.

> (Fallman, 2007, p. 305)

From our standpoint, the UI is an example of a complex language. Consequently, we can apply different UI language components such as: discrete elements, interaction sentences, narration, rhetorical tropes, and patterns. By analyzing the individual statements, we can follow an entire argumentation constructed with the help of the different UI elements. A simple way of doing this is transcribing the "interaction sentences," UI language components (Section 3.1), that the user encounters while performing a certain task. The interaction sentences can be analyzed further in terms of what goals the designers have and what assumptions they have about his or her users. By exploring different parts of the system through the UI, we can extract the inherent values. We argue that when the UI follows the structure of natural language, it both behaves more user-friendly and conveys the designer's intent more effectively.

According to Sengers (2010), HCI rests upon the modernist tradition, which follow these ideological themes: "Three themes are key to the way I am framing modernism here: (1) faith in technoscientific reasoning and expert knowledge as a way to organize our lives; (2) orientation around means-end thinking, maximizing efficiency and exerting control as fundamental ground principles to optimize everyday processes; and (3) closed-world thinking" (Ibid., p. 4).

The above themes can be noticed in the shift towards computation and software, as Bauman (2000) points out. Since its inception in the nineteenth century, the (western) modernist tradition of ideology orbits around five main concepts: emancipation (on a personal as well as social level), individuality (liberal ideology), time/space (fear of the stranger), work (with its emphasis on productivity), and community (nationalism, unity). We shift from "heavy" and "solid," hardware-focused modernity to a "light" and "liquid," software-based modernity (Ibid.). In order to tackle this problem, a shift of analysis towards this liquid phase is needed. We argue that software—both on the personal (user) and social (society) level—should be regarded as a driving force, a catalyst, for a certain type of behavior.

The term "ideology" was coined by Tracy not only as "the science of the formation of our ideas" (Tracy, 2009, p. 17), but also as their expression, and combination (Ibid., p. 1). By this definition, Tracy tries to restore the concept of logics as it was understood in ancient history. For our purposes, we understand ideology as:

[A] logically coherent system of symbols which, within a more or less sophisticated conception of history, links the cognitive and evaluative perception of one's social condition—especially its prospects for the future—to a program of collective action for the maintenance, alteration or transformation of society.

(Mullins, 1972, p. 150)

What criteria should we then use to recognize ideologies and analyze them further? Again, according to Mullins, these components are: cognitive power, evaluative power, action-orientation, and logical coherence (Ibid.). By cognitive power (Section 2.3.1), he means the "cognition and retention of information" (Ibid.), when we identify and symbolize our recurrent experience. After having done this cognitive process we can simplify, order, and abstract it for making choices between information, e.g., on different causal forces. The evaluative power (Section 2.3.2) is then based on this understanding of information. Political ideology, "incorporates evaluations of what is conceived" and can anticipate "possible events and conditions" (Ibid.). The action-orientation (Section 2.3.3) is based on the power of the ideology to "communicate conditions, evaluations, ideals, and purposes among members of groups (...) and thereby facilitates the mobilization and direction of energies and resources for common political undertakings" (Ibid.). Finally, the logical coherence (Section 2.3.4) or consistency between various ideology components means, "the ideology must 'make sense' and not result in logical absurdities" (Ibid.).

As the word suggests, "ideology" is related to ideas. On this level, it is needed to focus on the relation between UI and image. As Mitchell put it,

The concept of ideology is grounded, as the word suggests, in the notion of mental entities or "ideas" that provide the materials for thought. Insofar as these ideas are understood as images—as pictorial, graphic signs imprinted or projected on the medium of consciousness—then ideology, the science of ideas, is really an iconology, a theory of imagery.

(Mitchell, 1986, p. 164)

FIGURE 2.1 Soviet poster dedicated to the 5th anniversary of the October Revolution and IV Congress of the Communist International. The image contains all of the four ideology components defined by Mullins, 1972. Source: `http://commons.wikimedia.org/wiki/File:CominternIV.jpg`, cit. 2009-05-02.

"Ideology, then, which begins historically as an iconoclastic 'science of ideas' designed to overturn 'idols of the mind,' winds up being characterized as itself a new form of idolatry—ideolatry" (Ibid.). Thus, it is important to analyze the visual plane, (together with metaphors, mental models, navigation, interaction)[1] of UIs, where HCI ideologies take the most recognizable shape (for a visual example of ideology components, see Figure 2.1).

Currently, in the context of ICT, ideology comes to us from a rather unexpected direction. As Galloway (2008) points out, citing Althusser, ideology that was, "traditionally defined as an 'imaginary relationship to real conditions' (Althusser)" (Ibid., p. 953) has been superseded by simulation. He understands simulation as an "imaginary relationship to ideological conditions." In short, "ideology gets modeled in software" (Ibid.). Therefore, software serves as a prime example of current ideologies acting on us according to all the four criteria.

2.3.1 COGNITIVE POWER

Software models ideology and makes ideology visible through the way software works. This reflects the cognitive power of ideology by relating to the underlying hardware in a specific way:

---

[1] cf. Marcus, A. "Integrated information systems: A professional field for information designers." *Information Design Journal* 17:1, 4–21, 2009.

In a formal sense computers understood as comprising software and hardware are ideology machines. They fulfill almost every formal definition of ideology we have [...]. Software, or perhaps more precisely operating systems, offer us an imaginary relationship to our hardware: they do not represent transistors but rather desktops and recycling bins. Software produces "users."

(Chun, 2004, p. 43)[2]

Software creates both a relation with hardware as well as with users. Hardware is what the user encounters first, although the focus is then shifted to the software and the UI as a whole. UI is regarded as an entrance into a simulated world, but UI also forms a media layer between the "real" world and the user. "The doorway/window/threshold definition is so prevalent today that interfaces are often taken to be synonymous with media themselves" (Galloway, 2008, p. 936). An even more poignant definition relates the UI more tightly to the effect it has on the interacting users:

The interface is this state of "being on the boundary." It is that moment where one significant material is understood as distinct from another significant material. In other words, an interface is not a thing; an interface is always an effect. It is always a process or a translation.

(Galloway, 2008, p. 939)

The UI works thus not only on a semiotic level by differentiating symbols, but also on a psychological level when it creates relations and effects. For the UI to be effective and enjoyable, it is important to work "as a 'mirror' depicting the user's self-image, not only a 'window' looking into a world of content (...)" (Marcus, 1998, p. 53).

The differentiation work of the UI done between the user and users' self-image leads us to think about the UI in the terms of an active self-organizing entity. This notion is close to what Derrida (1993) called "difference." Following Derrida's argumentation, the UI presents a different idea from the original one (or content) just by the way it is mediated. Thus, different media can go only as far as their structure permits. The medium of text can express other things than speech (e.g., Derrida's example of difference vs. difference, both of which are read the same); the medium of image can express other things than text, etc. The medium of the UI thus expresses its content differently.

The primacy of text for Derrida is something we can also observe very well in software. Software can go past the interacting subject, which is in contrast with the UI, which is bound in the subject/object relation (Derrida, 1993) simply because it requires a user. And because the action is done through the UI, the UI privileges the content it presents. In this way, the UI not only tells us how to read a certain idea but can also preselect for us which ideas we can possibly read. Winograd and Flores state that: "Computers have an especially large scope, for they are machines that work with language. By using them, we join a discourse set up in the limits made by programmers" (Winograd and Flores, 1987, p. 178).[3] Each UI presupposes a certain

---

[2] cf. Galloway, 2008, p. 953—"The computer is the ultimate ethical machine. It has no actual relation with ideology in any proper sense of the term, only a virtual one."
[3] cf. Derrida, 1993.

paradigm of use which is not always visualized. When built correctly, the UI lets us see only what has to be seen. The UI itself stands on a certain HCI ideology. It defines relations which are to be made. The relations made by ideology are political inasmuch as they are social.

While the prevalent UI definition is connected with a gateway as a passage into another world, beyond the entrance this world is structured by another narrative. By analyzing the narrative, we can gain a better insight into the UI structure and the underlying HCI ideology:

> In temporal terms, narrative is about what already happened while simulation is about what could happen. Because of its static essence, narrative has been used by our culture to make statements. (. . .) The potential of simulation is not as a conveyor of values, but as a way to explore the mechanics of dynamic systems.
>
> (Frasca, 2004, p. 86)

2.3.2 EVALUATIVE POWER

The user is presented with information "designed to program the spectators of techno-images to behave in a specific way and this in turn serves as a feedback to the programs calculating these techno-images" (Flusser, 1995). Here, techno-images are computer-generated images in Flusser's theory. Let's take the example of Google Earth (see Figure 2.2), which builds up on our

> belief that a map covers a concrete phenomenon, my "taking for true." The function of my map—and of all the techno-images—lies in the effort to impose on me a programmed idea of a concrete world, thus to program my cognition and evaluation of the world and all of my acting in the world.
>
> (Flusser, 1995)

Therefore, for building new UIs, we ought to deconstruct the present ones and uncover their design/intent. Winograd and Flores also suggest this by stating that: "design is the interaction between understanding a creation. . . [We therefore] need to set up a theoretical framework not to watch how the devices operate, but what they cause" (Winograd and Flores, 1987, p. 53). This is frequently the only way to understand new UIs in a situation when we do not have a suitable interpretation key—we do not know their code. It is, in a way, something like "reverse engineering" known from computer science.

2.3.3 ACTION ORIENTATION

In order to use the UI, different languages are present in the form of action paradigms. "Action paradigms define a set of instructions that are available at any given moment. The paradigms offered by the system should match those the user needs so that she's not forced to perform an action she didn't intend" (Andersen, 1997, p. 91). For example, take the interaction game for putting the computer to sleep in Microsoft Windows XP. Here the user has to first click on Start, then Shut Down, only then is he or she presented with the intended Sleep button. Thus, for putting the computer to

FIGURE 2.2 A Google Earth screenshot depicting the historical center of Prague in a very realistic way. Map data ©2008 Google, ©2014 GEODIS Brno, SIO, NOAA, U.S. Navy, NGA, GEBCO. Google and the Google logo are registered trademarks of Google Inc., used with permission.

sleep, we have to choose from UI language components that are in conflict with our intent. Even when something does not work as expected, we can gather interesting data out of it. When we interpret a connection between a UI sign and a proposed function, this mental connection is what forms our image of the system. "Systems work because they don't work. Non-functionality remains essential for functionality" (Galloway, 2008, p. 931). Similar oppositions build the interaction space as:

> . . . the "choices" operating systems offer limit the visible and the invisible, the imaginable and the unimaginable. You are not, however, aware of software's constant constriction and interpellation (also known as its "user-friendliness"), unless you find yourself frustrated with its defaults (. . .).

(Chun, 2004, p. 43)

The action-orientation of ideology also works when the medialization (i.e., how the content is presented to the user) is not credible. In such a case, however, the medialization works the other way around: it influences our intent according to what can be medialized. However, for a UI to be effective, it should be both credible and familiar: "Designing for familiarity is crucial when trying to persuade people to behave in unfamiliar ways" (Wai, 2007, p. 99).

The user actuates the computer (or apparatus) to use it together with their technical imagination to create something but, paradoxically, one of the computer functions is the user's intent (Flusser, 2001, p. 24). This is so because the apparatus is predisposed only for some type of code and program cycles. As Bogost (2007) says: "Software establishes rules of execution, tasks and actions that can and cannot be performed"

(Ibid., p. 4). Therefore, for the designer's intention to be fulfilled, he or she can intend only what is achievable. Only by using a specific apparatus for the chosen job can the designers' intent be fulfilled: "The freedom of decision of pressing a button with one's fingertips turns out to be a programmed freedom. A choice of prescribed possibilities. I choose according to the regulations. . . " (Flusser, 1999, p. 93). Such freedom leads to the illusion of nearly unconfined freedom, however, our interactions are latently directed to a certain goal. This freedom leads us to take over the thinking of the designer.

The above-mentioned "programmed freedom" is closely connected with procedures as sequences of action. Again, with Bogost (2007), "[p]rocedures are sometimes related to ideology; they can cloud our ability to see other ways of thinking (. . .)" (Ibid., p. 3).

We can take the action-orientation element of ideology as a form of rhetoric. This view is further discussed in Chapter 9. Since Aristotle, rhetoric was used in different media to state arguments of the designer in order to make the audience believe in the presented reality. Persuasion as a technique has made its way into ICTs and has even been transformed into a tool. Fogg (2003) defines a persuasive technology tool as "an interactive product designed to change attitudes or behaviors or both by making a desired outcome easier to achieve" (Ibid., p. 32).

In the ICT environment, the persuasive tools are supported by the inner workings of software, as we have stated above. These workings, based on procedures, help to get predefined arguments to the users. Bogost (2007) calls it "procedural rhetoric." "Procedural rhetoric is a technique for making arguments with computational systems and for unpacking computational arguments others have created" (Ibid., pp. 2–3).

A specific characteristic of procedural rhetoric is that it does not build arguments using techno-images, but "through the authorship of rules of behavior, the construction of dynamic models" (Bogost, 2007, p. 29). Therefore, procedural rhetoric works in the space of medialization between design/intent and design/form. In such a manner, it is close to a "UI grammar" (Brejcha and Marcus, 2013), where language plays the part of a rule system. In the system, the UI designer establishes grammar rules (syntax) for the combination of its elements. The manner in which UIs are built is governed by a set of rules given by the designer, for example, every UI produced can follow a different intrinsic UI grammar. The choice of elements is then subject to the pragmatics of the entire UI.

2.3.4 Logical coherence

Mullins suggests that ideology should be coherent, that is, syntagmatic rather than paradigmatic, since it needs to help create a seamless experience. From the perspective of internal connectedness, design fulfills the same function as art, technics, and machines for they manipulate and try to master the original state of things, nature (Flusser, 1999). As Flusser (Ibid., p. 19) continues: "This is the design that is the basis of all culture: to deceive nature by means of technology, to replace what is natural with what is artificial and build a machine out of which there comes a god who is ourselves."

As we implied above, UIs are intersubjective media. Winograd and Flores (1987, p. 169) support this by saying, that "by producing tools we design new conversations and new relations." Therefore, the things for use mediate human relationships. And on this level signs (i.e., elements of representation) are also created (Schütz, 1973, p. 148). The design thus sets forth human relations. In a lot of cases this is done with a certain goal, as it is in social web projects, such as Facebook. It must be clear, however, that in most cases this is done inadvertently. Here, the agent is no more the designer, but the system of codification and medialization, determined by technical devices, above which the creator has no power any longer. What is important here is that the ideology perpetuates itself beyond the human reach.

> Programmers aren't the important elements for the functioning of techno-images, but the structures of apparatuses they produce. Techno-images are imperativistic not because they are used by some ideologists to manipulate the society, but because they are a projection of such a pixel universe, that pretends to present the world pixel by pixel. For this imperativistic, "imperialistic" nature of techno-images not the human being, but an artificial plotter, artificial intelligence, automatism of apparatuses is in charge, and has become independent from the human.

> (Flusser, 1995)

In the above quote, what is imperativistic is the constructed artificial world that forces us to take it for reality. What is imperialistic is the tendency of the producers (or even the producing automata of techno-images themselves) to colonize the semiotic space with signs (techno-images) referring to other techno-images and leaving out all the rest. Such a tendency is supported by a number of ideologies embedded in the UI.

What are then the emerging HCI ideologies present in the UI? In the field of UI design, different instances of ideology are present. So far, one of the most prominent is the ideology of hypertext (Bush, 1945). As Nielsen states, "[hypertext] makes individual users the masters of the content and lets them access and manipulate it in any way they please" (Nielsen, 2004). This user-empowering approach is contrasted by choice-obfuscation (e.g., when navigation links are not readily visible) or even user oppression (when user choice is limited or eliminated, e.g., in splash screens or ads) (Ibid.). Currently, the semantic space of UI ideology is somewhat centered around the terms "simple, fast, intuitive, social, minimal, choice, useful, fun," as a series of interviews with web designers suggest (Chang, 2006). Relating to the understanding presented above, perhaps the leading ideologies are: the semantic web, Open Source movement, the hacker ethic (Levy, 1984) and Wikipedia, all of which follow the empowering principle.

Another important ideology is the ideology of ease. Dilger (2000) presents the ideology of ease, which dissects users into the "computer illiterate" and "techies" and suggests that this "will ensure that the historical boundaries of gender, race and class are reproduced in computing practices for years to come." By ideologies, he means the "frameworks of thinking and calculation about the world—the 'ideas' that people use to figure out how the social world works, what their place is in it, and what they ought to do" (according to Dilger's reading of Hall, 1985). It is important to mention that in the same manner HCI/UX practices reproduce also cultural, and age boundaries, and possibly others. This agrees with Mullins' view, since the way the world works refers to cognitive and evaluative power, people's place in it, and what they ought to do then refers to action-orientation. Dilger states, that ease is gendered, which is to be seen in the connotation of an "easy" to use computer system as feminine. Ease has a different meaning in connection to work and leisure, during the former it has to be supported by the system, during the latter a certain difficulty could be desirable, e.g., in chess. At work, moreover, a task may not seem worthwhile if it does not seem easy. Pictures may furthermore seem easier to understand than text, which is supported by various media, such as television or comics. The notion of speed is also connected to anything which would be labeled as easy including learning. Finally, the gain of ease is matched by a loss in choice, security, privacy, or health (Ibid.).

Some of the HCI ideologies may even have a more pronounced impact on the user's behavior through the use of social cues and persuasion techniques. In the mobile context this has been carried out by Aaron Marcus and Associates, Inc., in a series of mobile applications. The applications may lead the users to reduce their ecological footprint (Marcus and Jean, 2009), reduce weight, improve dietary behavior (Marcus, 2011), manage wealth after retirement (Marcus, 2012a), or share memories and family wisdom (Marcus, 2012b), among others.

2.4   UI LANGUAGES

As stated earlier, we assume that HCI takes place between different actors (users, systems, designers) in a specific setting or paradigm. The semiotics of interaction is closely related to language as a system of signs. The semiotics of interaction is by definition time-based and the same holds true for language. Because of this, we may find some interesting parallels. The HCI/UI designer establishes grammar rules (syntax) for the combination of its elements. UIs are built from different components (metaphors, mental models, navigation, interaction, appearance) (Marcus, 2002). The manner in which UIs are built is governed by a set of rules given by the designer, for example, every UI produced can follow a different intrinsic language grammar. The choice of elements is then subject to the goal (pragmatics) of the entire UI (see Garrett, 2002). Therefore, we should structure the UI language according to the actors and audience we want to address.

In order to better grasp the domain of language communication (in our case, how the UI embodies language concepts), the semiotics of interaction borrows some concepts from linguistics (and partly from cognitive science, especially mental models). Perhaps the most prominent in this respect is the linguistic model of language, which describes specific orders of subject, verb, adverb, and object in a sentence. Interestingly, Indo-European languages give prominence to contrasts to build a grammar logic (substantive and verb) (Whorf, 2012, p. 309), in that they are binary. In contrast, "Nootka has no parts of speech; the simplest utterance is a sentence," and "[l]ong sentences are sentences of sentences" (Ibid., p. 310). Because of these differences, we will focus first on the English language. All the combinations exist, but the prevailing model is subject-object-verb, or SOV (see Figure 2.3) (Dryer, 2008), which is mimicked in graphical UIs (GUIs). The user first must select the object (e.g., a file) then the corresponding action (e.g., move to trash). A different sequence prevails in

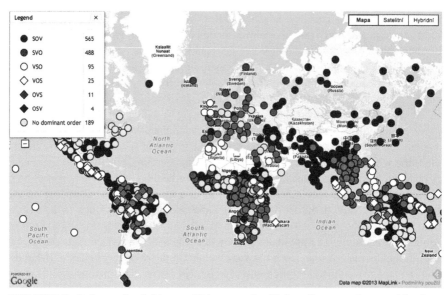

FIGURE 2.3 Order of Subject, Object and Verb. Source: `http://wals. info/feature/81A?z1=2435&v2=cd00&v1=c00d&s=20&v3=cff0&v4= dff0&v5=dd00&v6=d00d&v7=cccc&z5=2989&z6=2996&z7=2811&z4= 2975&z2=2512&z3=2905&tg_format=map&lat=31.653381399664&lng= 8.7890625&z=2&t=m`, cit. 2013-05-01. Reprinted with permission from Matthew Dryer and Martin Haspelmath.

the command-line (user) interfaces (CLIs). The model used is subject-verb-object, or SVO, which follows the natural word order of the English language.

The most important part of such a linguistic model is, however, the verbal phrase, the subject being the user interacting with the computer. Or, as Nadin put it, "as soon as we interpret a sign, we become part of it for the time of that interpretation" (Nadin, 1988, p. 271). The user, by interpreting the sign, connects to the sign interpretant to decode the intended meaning.

As mentioned before, the differences among linguistic models of different languages also dwell in the order of adverbs (e.g., in English, manner, place, frequency, time, purpose [Capital Community College Foundation, 2005]). We suggest that following the natural adverb order in the UI would benefit the user's mental model and thus the overall usability. The adverb order can structure the sequence (narration) of the UI interaction, together with the distinction between theme/rheme, old/new information, and background/focus. The adverb order can be seen, for example, in a wizard-based discourse (when we are booking a flight on the Web, we are presented with choices: From what origin? To what destination? When? How many travelers? etc.) (see Figure 2.4).

We argue the user's native language determines his or her mental models (i.e., typical way of thinking). Here, we refer to the linguistic relativity theory (Sapir–Whorf hypothesis), which maintains that each language can perpetuate or "predetermine for us certain modes of observation and interpretation" (Sapir, 1949, p. 11; see also Nisbett, 2004). In other words, "users of markedly different grammars are pointed

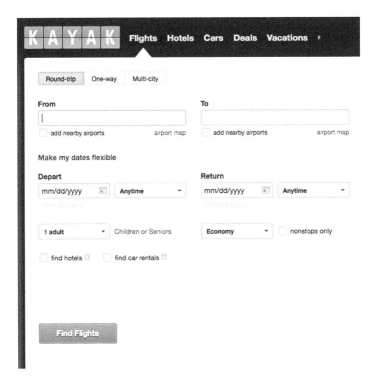

FIGURE 2.4 Kayak air ticket finder. Source: `http://www.kayak.com`, cit. 2013-05-18. KAYAK screenshot reprinted with permission from KAYAK Software Corporation.

by their grammars toward different types of observations and different evaluations of externally similar acts of observation, and hence are not equivalent as observers but must arrive at somewhat different views of the world" (Whorf, 2012, pp. 282–283).

Or, as Halliday (1985) formulated, "[g]rammar goes beyond formal rules of correctness. It is a means of representing patterns of experience [...]. It enables human beings to build a mental picture of reality, to make sense of their experience of what goes on around them and inside them" (Ibid., p. 101). By expressing in a different language, we accent different objects and different experiences, which enables us to get different insights into the world we live in. This change of perspective is analogous, when we shift from the traditional system of reference of HCI based on computer science to a semiotic and linguistic system, which has native tools to work with meaning.

"Meanings [however] belong to culture, rather than to specific semiotic modes. And the way meanings are mapped across different semiotic modes, the way some things can, for instance, be 'said' either visually or verbally, others only visually, again others only verbally, is also culturally and historically specific. [...] what is expressed in language through the choice between different word classes and clause structures, may, in visual communication, be expressed through the choice between different uses of colour or different compositional structures. And this will affect meaning. Expressing something verbally or visually makes a difference" (Kress and

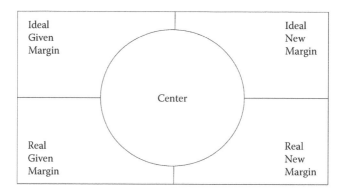

FIGURE 2.5 Spatial map for assessing the layout semiotics according to Kress and van Leeuwen.

Van Leeuwen, 2006, p. 2). Therefore, there is a difference between a speech act and an image act.

Kress and Van Leeuwen (2006) describe the function of the image act: "The producer uses the image to do something to the viewer. It is for this reason we have called this kind of image a 'demand,' following Halliday (1985): the participant's gaze (and the gesture, if present) demands something from the viewer, demands that the viewer enter into some kind of imaginary relation with him or her" (Kress and Van Leeuwen, 2006, pp. 117–118). Another kind of image is an offer: "it 'offers' the represented participants to the viewer as items of information, objects of contemplation, impersonally, as though they were specimens in a display case" (Ibid., p. 119). Both concepts can be related to speech acts. Kress and Van Leeuwen suggest, that despite the similarities "it would seem that 'image acts' do not work in the same way as speech acts. When images 'offer,' the primarily offer information. [. . .] When images 'demand,' they demand [. . .] the 'goods-and-services' that realize a particular social relation. [. . .] There is no image act for every speech act" (Ibid., p. 123). The last sentence reaffirms the distinct characters of the two kinds of acts.

There are other examples showing the distinctive features of speech acts and image acts. "In the depiction of humans (and animals), 'involvement' and 'detachment' can interact with 'demand' and 'offer' in complex ways" (Kress and Van Leewuen, 2006, p. 138). Involvement is supported by showing the object in the eye's height and facing directly the viewer, as is the case in most graphical UI icons. "How is 'involvement' realized in language? [. . .] the two systems, the visual system of horizontal angle and the linguistic system of possessive pronouns, differ in many ways" (Ibid., p. 139). However, the visual system can be related to the linguistic system through the meaning of space.

For example, there are certain places on the layout of a newspaper, a computer screen, etc., that can convey a specific implied meaning, when combined with objects, or composition elements. In Western culture (see Figure 2.5) "the elements placed on the left are presented as Given, the elements placed on the right as New. For something to be Given means that it is presented as something the viewer already knows, as a familiar and agreed-upon point of departure for the message. For something to be New means that it is presented as something which is not yet known, or perhaps not yet agreed upon by the viewer, hence as something to which the viewer must pay special attention" (Kress and Van Leewuen, 2006, p. 181). Other cultures can share these spatial meanings, or can have different ones, as our research suggests (see Part II, "Culture of Interaction").

The cultural differences are present not only on the spatial plane, but also on the temporal plane, which is manifested, for example, by intonation. "As in visual communication, the structure of a 'tone group,' an intonational phrase, is not a constituent structure, with strong framing between elements, but a gradual, wave-like movement from left to right (or, rather, from 'before' to 'after,' since in language we are dealing with temporally integrated texts), and it is realized by intonation. Intonation creates two peaks of salience within each 'tone group'—one at the beginning of the group, and another, the major one [. . .], as the culmination of the New, at the end. [. . .] In other words, there is a close similarity between sequential information structure in language and horizontal structure in visual composition, and this attests to the existence of deeper, more abstract coding orientations which find their expression differently in different semiotic modes. Such coding orientations are culturally specific, certainly where the horizontal dimension is concerned" (Kress and Van Leeuwen, 2006, p. 181). The horizontal structure of composition is coupled with the vertical composition.

In the vertical structure, "what has been placed on the top is presented as the Ideal, and what has been placed at the bottom is put forward as the Real. For something to ideal means that it is presented as the idealized or generalized essence of the information, hence also its [. . .] most salient part. The Real is then opposed to this in that it presents more specific information (e.g., details), more 'down-to-earth' information (e.g., photographs as documentary evidence, or maps or charts), or more practical information [. . .]" (Kress and Van Leewuen, 2006, p. 187). Again, the vertical structure is also culture-dependent.

An important complement to the horizontal and vertical axis is the center vs. margin. "For something to be presented as Center means that it is presented as the nucleus of the information to which all other elements are in some sense subservient. The Margins are these ancillary, dependent elements" (Kress and Van Leeuwen, 2006, p. 196). All of the above structures can work together in concert. "One common mode of combining Given and New with Center and Margin is the triptych" (Ibid., p. 197). "The structure of the triptych, then, can be either a simple and symmetrical Margin-Center-Margin structure or a polarized structure in which the Center acts as a Mediator between Given and New or between Ideal and Real [. . .]. Though spoken English has its own Given-New structure, this is not the case with the Ideal-Real and the Center-Margin structures. This not to say that the meanings these structures express cannot, in some form, be expressed in language, but rather that they are more readily and frequently expressed visually, and that language, unlike visual communication, has not developed 'grammatical' forms to express them" (Ibid., p. 199).

The above-mentioned structures provide the building blocks for employing UI languages in the HCI/UI design. These design languages share an analogy to the natural language, but are "the basis for how we create and interact with things in the world" (Rheinfrank and Evenson, 1996, p. 65). And further, "design languages are used to design objects that express what the objects are, what they do, how they are to be used, and how they contribute to experience" (Ibid., p. 68). According to Sutherland

(1966) there are two principal languages involved in HCI: "One is the language of display by which the machine presents information regarding the state of its data and the options available for further action by the user. The other is the language of actions using input devices, by which the man relates his intended transformations of machine-stored data with references to objects in the displayed picture" (cited from Foley and Wallace, 1974, p. 465). We shall explain them in the following sections.

2.4.1 VISIBLE LANGUAGE

The language of display is in most of the cases ruled by a visible grammar, although different grammars for different perceptions (such as auditory, tactile, oral, olfactory) can still be used. In our case we focus primarily on the interaction grammar, as well as the visible grammar, that starts the interaction by providing a visual narrative that could be followed. The language of display corresponds therefore to a visible language.

By visible language we mean a systematic language of expression conveying specific information, that can be translated from one kind of language to another (Marcus, 2003b; see also Moles, 1966). The visual language, on the other hand, refers to loose visual means for conveying emotions or general concept, as used by artists. Currently, several authors analyze visible languages (Roam, 2011; Kress and van Leeuwen, 2006; Engelhardt, 2002; Horn, 1998; Narayanan and Hübscher, 1998; McCloud, 1994; Bertin, 2011) which could help us methodologically in the area of interaction and communication design. The visible language represents numbers, nouns, and verbs (forming a narration) (see Tufte, 1983, 1991, 1997).

2.4.2 INTERACTION LANGUAGE

The language of actions is based on user input, and it thus corresponds to the inter-action language. Based on our decision to act (or refrain from acting) on an object, we start an interaction that modifies the current screen. The interaction language is based on a grammar, that connects all of the semiotic dimensions mentioned earlier in Chapter 2, "Semiotic Foundations. . . ."

Let us remind Peirce's definition of semiotics related to syntax: "Semeiotic grammar is concerned with determining the formal conditions for signs as such (CP 1.444)" (cited from Liszka, 1996, p. 18). The grammar sets rules on what can be chained in a cause-result interaction unit, how it can be chained, and on what purpose. Because of these rules, the grammar of interaction can establish what can be expected based on the metaphor, mental model, navigation, interaction, and appearance in use on the working screen. Therefore, the interaction grammar can support the consistency of all the UI components.

Consistency helps to build expectations, which in turn establishes a set of constraints for the interaction and communication design. Constraints should be used, as Laurel (1993, p. 105) suggests: "without shrinking our perceived range of freedom of action: Constraints should limit, not what we can do, but what we are likely to think of doing." For Norman (2002) these constraints are physical, semantic, cultural, and logical. The physical constraints "are closely related to real affordances: For example, it is not possible to move the cursor outside the screen [. . .]" (Norman, 1999, p. 40). Logical constraints "use reasoning to determine the alternatives. Thus, if we ask the user to click on five locations and only four are immediately visible, the person knows, logically, that there is one location off the screen" (Ibid., p. 40). And finally, the cultural constraints "are conventions shared by a cultural group. [. . .] The choice of action is arbitrary: there is nothing inherent in the devices or design that requires the system to act in this way" (Ibid., p. 41). Only the cultural constraints require a higher level of interpretation and a broader context, which makes them self-explicative.

By exploiting the user's learned cultural background, on the other hand, we can propose a much richer (and even more effective) interaction. In order to do so, it is important to be able to analyze the HCI/UI from a cultural perspective. Towards this end we shall compare two methods for UI annotation and analysis to show their possible benefits in this process.

3 Semiotic Design and Evaluation Framework

3.1 UI LANGUAGE COMPONENTS

The visible and interaction language is expressed through UI components. On a general level, UI components concern how a human-computer system works by means of intrinsic metaphors, mental models, navigation, interaction, and appearance (Marcus, 2009, p. 9):

- Metaphors: Essential concepts in words, images, sounds, touch
- Mental models: Organization of data, functions, tasks, roles of people at work or play, static or mobile
- Navigation: Movement through mental models via windows, dialogue boxes, buttons, links, etc.
- Interaction: Input/output techniques, feedback, overall behavior of systems and people
- Appearance: Visual-verbal, acoustic, tactile qualities

Looking at the UI components from the language perspective, we can structure them organically to create a UI grammar. UI grammar is composed of basic elements: interaction sentence, interaction games, rhetorical tropes, interaction phases, and patterns. The grammar elements concern both the noun and verb phrase of a sentence. Discrete elements are the smallest elements to have a meaning. The interaction sentence is a meaningful unit describing a task in a user's interaction. A set of interaction sentences with the same goal forms an interaction game. The narrative in UI is made both by the designer's metacommunication and the temporal aspect of perceiving UI elements. Rhetorical tropes are devices of persuasion and emphasis, often presented as metaphors. Patterns are typical configurations of UI language components in different settings. From the defined semiotic and UI language principles, we extracted a set of heuristics that could be used as an UI glossary both for evaluation and design.

3.1.1 DISCRETE ELEMENTS

Discrete elements are the smallest elements to have a meaning. In linguistics they are called morphemes, or sememes in semantics. Morphemes modify the lexemes, for example in the lexeme "mouse-down" the morpheme is "-down." The smallest elements without a semantic meaning are called graphemes. In written language they would include letters of the alphabet or strokes in Chinese characters; in the UI they are present as affordances communicating attributes, such as shapes, shadows, or colors (see Gibson, 1977; Norman, 2002).

Lexemes and graphemes could be used to capture the user's interaction, but for the actual analysis we should work with higher-level elements. For example, Richards (1984) believes that "there seems to be little profit in using such items as an individual dot or line as a unit of analysis. If we are going to use linguistics as a model, then what is needed for present purposes is not the pictorial equivalent of a phoneme or morpheme but something closer to a noun phrase" (cited from Engelhardt, 2007, p. 27). However, some of these elements are useful to provide additional information, such as color. In such a case "signifiers—and colours are signifiers, not signs—carry a set of affordances from which sign-makers and interpreters select according to their communicative needs and interests in a given context. [. . .] First there is association, or provenance—the question of 'where the colour comes from,' where we have seen it before.' [. . .] The second type of affordance is that of the 'distinctive features' of colour" (Kress and Van Leeuwen, 2006, pp. 232–233).

In the interaction syntax one can identify "lexemes," or interaction "words" present in an interaction dictionary, (Hjelmslev, 1961, cited from Andersen, 1997, p. 256), such as mouse-up, mouse-down, mouse-drag. A "word" in tangible UIs, or TUIs, can be "touch." One or more lexemes form lexical items, or actions, like "pointing with a mouse" (Payne, 1991, p. 136), which build up a vocabulary of interaction language.

The standardized vocabulary (lexical items) can be UI widgets (we refer to the primitives, like links, buttons, icons, pop-up controls, etc., but not to more complex forms, like Google widgets or Apple gadgets, which are UIs in themselves). These basic elements can consist of texts, pictures, sounds, animations, etc. To exclude too much complexity the widgets must fall in the category of noun-phrase (e.g., selected radio-button), or verb-phrase (e.g., type of action, which could/should be performed).

Identifying the discrete elements can help us assess the economy of elements used. It is important to be able to "form an infinite number of meaningful combinations using a small number of low-level units (offering economy and power). The infinite use of finite elements is a feature, which in relation to media in general has been referred to as 'semiotic economy'" (Chandler, 2001). In interaction and communication design we should seek to minimize the elements used, thus maximizing the potentional economical gain in clarity and simplicity of the UI.

The lexical items, in turn, construct an interaction "phrase." The interaction phrase consists of noun phrases (involving the subject and object) and/or verb phrases (involving a verb, which provides more information about the subject). The interaction phrases together form an interaction sentence.

3.1.2 INTERACTION SENTENCES

The interaction sentence is a meaningful unit describing a task in the user's interaction. For example, in a drill-down menu, the user selects the category he or she needs to modify, then selects the parameters to set, and then presses the "OK" button. The visual hierarchy promotes a certain reading (and interpretation) sequence of the UI elements.

The interaction sentence is subject to the visual elements, which, through visible grammar, can lead the user to act. Each user-driven event in the UI is constituted by (a) an intention, and (b) an action. The intention is manifested by mouse-over (pointing), selecting a widget or command. The action is manifested by mouse-down (click) on a "go" button or menu item, releasing the mouse over the command, or dragging. As Marcus (2003a) holds, "All widgets are about some selection/indication of intent" (Ibid., 4.4.8). Any visual element in the UI can thus start a sentence and continue a dialogue with the user. According to Foley and Wallace the essential features of the sentence structure are: "indivisible, complete thought; unbroken actions; a well-defined 'home state'; regularity of pattern" (Foley and Wallace, 1974, p. 465). They provide examples of such sentences, such as:

> Draw a line from this point to that one. Apply this constraint to that object. Rotate this object about that axis by the following amount.

(Foley and Wallace, 1974)

The sentences should match user's natural language (mental model) as closely as possible for the interaction to be natural ("intuitive") and effective. The level of clarity can be assessed during evaluation in the form of think-aloud, when the user utterances are analyzed and categorized in order of relevance and priority (i.e., how well, if at all, the designer's intended meaning is transferred to the user, as explained by de Souza, 2005), or checked for consistency (syntax) and meaningfulness (semantics). Although the main focus here is on visible and interaction syntax, we should seek the best possible syntax-semantics alignment.

Interaction sentences, in order to match user's mental models and allow for a natural flow of interaction, should follow "a number of syntactic principles of naturalness for action sequences" (Foley and Wallace, 1974, p. 465) besides the aforementioned sentence structures, also "visual continuity, tactile continuity, and contextual continuity" (Ibid.).

According to Brandt (1993, p. 138), "the flow becomes an active, transcendental principle, an instance that carries a kind of general state of belief that cannot be stated explicitly, but only shows its effects in the work that the semiotic flow yields at the separate stations."

From a semiotic analysis perspective, an interesting possibility is extracting the possible interaction sentences from the UI, and by doing so, being able to evaluate the UI. This idea is based on the concept, that "language provides facilities for controlling the information structure of the sentence: theme/ rheme, old/new information, and background/focus" (Andersen, 2001, p. 6). Andersen even believes that a series of sentences can be translated into a UI form, thus helping to create the UI.

There are already different kinds of user-interaction notations that have been used since the 1980s (e.g., Backhus-Naur Format annotation, or BNF; Task-Action Grammar, or TAG—Payne and Green, 1986; eTAG (de Haan, 2000); notation based on object-oriented programming—Andersen, 1997; adapted Kats and Fodor model from Eco (1979) in O'Neill (2002)). These, however, produce a pseudoprogramming code, which often provides too much detail, is hard to read, and does not scale well for more complex UIs.

In the context of HCI, a similar concept to interaction sentences is a scenario. According to Carroll (2000, p. 46), "[s]cenarios are stories—stories about people and their activities." The interaction sentences share with the scenarios their incompleteness. As Carroll (2000) says, "Scenario descriptions are fundamentally heuristic;

indeed, they cannot be complete. For any nontrivial system, there is an infinity of possible usage scenarios. The incompleteness of scenario descriptions is an important property but not unique to scenarios. The only complete specification of a system is the system itself, and the only complete specification of the use of a system is an infinite log of its actual use, along with comprehensive cognitive and social analysis of that log" (Carroll 1990b cited from Carroll 2000, p. 255). Scenarios can be either simple or complex, can describe either a partial interaction, or a whole unit. As an example, we can cite (we separated the interaction sentences in lines):

> A person turned on a computer, the screen displayed a button labeled Start, the person used the mouse to select the button.

(Carroll, 2000, p. 14)

An accountant wishes to open a folder on the system desktop in order to access a memo on budgets. However, the folder is covered up by a budget spreadsheet that the accountant wishes to refer to while reading the memo. The spreadsheet is so large that it nearly fills the display. The accountant pauses for several seconds,

> resizes the spreadsheet, moves it in partially out of the display, opens the folder, opens the memo, resizes and repositions the memo, and continues working.

(Carroll, 2000, p. 46)

Scenarios can capture more data about user's postures, emotions, and thinking processes. While scenarios can provide important elements for describing a task in general, and accessibility through their informal character, they lack a clear structure, and are not necessarily focused on tasks to be carried out using a UI.

Use cases (Jacobson, 1992), on the other hand, emphasize well-defined and goal-oriented steps of user-computer interaction. These steps reflect the work done by all of the actors, that is, also of the computer system. Therefore, use cases work best for software development, but lack contextual data; that could prove useful for a semiotic analysis. With their orientation to a goal of interaction they are close to the interaction games discussed further. Use-case scenarios, which are employed in the context of use cases, focus on "one path through the use case" (Sharp et al., 2007, p. 510). This way, they are similar to the scenarios introduced by Carroll (2000). The advantage of use cases is, they can be used for complex, and nonlinear interactions.

A simpler way of capturing tasks provide Hierarchical Task Analysis (HTA) or GOMS (Goals, Operations, Methods, and Selection rules) for more detail (Sharp, et al., 2007, pp. 515–516). The task analysis lists all of the steps the user has to pass to achieve a goal, thus focusing on the user's context rather than the system processes. On the negative side, HTA is not suitable for complex or concurrent interactions.

Interaction sentences are constructed to honor the aforementioned principles of structure, clarity, and continuity, while allowing for narration details, and user-focus. Interaction games, on the other hand, are goal-oriented sets of interaction sentences. The succession of sentences, and creation of interaction games, is subject to the user's intention to achieve a goal. If the intended goal has been achieved, there is no need to continue with other interaction sentences.

We base our interaction games on the language game theory first introduced by Wittgenstein (1986). Here, we shall follow their application to interaction and communication design by Andersen, as follows.

> A language game can only be identified by the results the actions have in a given situation. A language game is a closed unit of interaction in the sense that when the last word has been said, nobody is required to continue the line of argument, since the goal has been reached or abandoned. A language game consists of sequences of verbal acts that mutually presuppose each other, and form a well-defined unit in relation to other acts, since they can be subsumed under the same purpose. (Holmqvist, 1986, p. 22; cited from Andersen, 1997, pp. 347–8; see also Wittgenstein, 1986)

On the other hand, "[v]irtually any meaningful sentence should be eligible for expression upon termination of another sentence" (Foley and Wallace, 1974, p. 465).

The interaction game (or simply "interaction") begins, when the user encounters the UI and needs to figure out which tasks to perform in order to achieve his or her goal. An objective might be to print out a specific document. The tasks (actions) to accomplish the goal are the following: searching for the document, opening it, and printing it. The end of the interaction game comes when the user evaluates the results and doesn't need to perform any other action to achieve his or her goal.

As we can see, the interaction game is closely tied to the user's interpretation of his or her actions. Transformed for the ends of UI interaction, we are (a) encountering a setting, (b) performing an action, and (c) seeing a result. This sequence accords with Norman's seven stages of action (Norman, 1986):

1. Forming the goal
2. Forming the intention
   — Gulf of execution —
3. Specifying an action
4. Executing the action
5. Perceiving the state of the world
6. Interpreting the state of the world
   — Gulf of evaluation —
7. Evaluating the outcome

(Norman, 1986, p. 47)

During the user interaction we can thus distinguish several phases. Here, we can build upon the thought, that "[w]e expect to have a beginning, a middle, and an end" (Laurel, 1993, p. xiii). This expectation is based on the drama tradition since Aristotle, who sets the following structure. The plot must be a whole, complete action. Such action is "that which has beginning, middle, and end. A beginning is that which is not itself necessarily after anything else, and which has naturally something else after it; an end is that which is naturally after something itself, either as its necessary or usual consequent, and with nothing else after it, and a middle, that which is by nature after one thing and has also another after it" (Barnes, 1984, p. 2321).

The interaction phases work both on the interaction sentence level and on the interaction game level. As such, the phases form a recurring pattern that is present in every interaction. More patterns are discussed in Section 3.1.5, "Patterns."

3.1.3 NARRATION

The narrative in UI is made both by the designer's metacommunication, and the temporal aspect of perceiving UI elements. The perception is guided through UI languages. This UI narration runs on two scales: the first one on the current screen (local), the second one through a whole sequence (global), a chained set of screens (or window states). If the designer's narration doesn't fully correspond to the user's understanding of the interaction with the UI, the user should at least be able to follow it along.

By *reading* the visible language of a screen we follow a certain path on screen according to what we want to accomplish, and thus read the seemingly helpful elements first. By interpreting the signs present on a given screen, we order them in a temporal manner in a sequence. This sequence takes on the form of a narration/storytelling layer of the UI. We call it the local narration.

The local narration is supported mostly by the appearance, metaphors, and mental model component of the UI. The local narration is constituted mainly by noun-phrases or descriptive sentences corresponding to a film shot ("Here is a revolver"). Because of this, the local narration is based on a static state of the screen, and on the UI components, that does not represent an action (as in the case of the interaction component). Here, this type of narration is articulated by the visible language and can form an argument or a statement.

Such narration is supported by the order of perception or appearance of the UI element and a goal (e.g., a motivation to search for information). The order of perception can be subject to the reading direction, size, shape, or color that would attract the eye to certain points first and thus creating a sequence (see Sutnar, 1961 for his psychology-based information design). The relation between elements on a single screen can be emphasized by distance, same/similar color, or line (for a detailed explanation see Bertin, 2011).

The sequence of appearance of an UI element can be controlled better by the designer, because it is not so much subject to the user's perception and interpretation. Moreover, as Eisenstein shows by the Kuleshov effect (Eisenstein, 1975; see also Metz, 1974), the viewer inadvertently relates two scenes coming one after the other and puts them into a narrative sequence. As Kress and Van Leeuwen (2006) shows us combinations of disconnected comics panels, or film shots that lead to different interpretation of the narrative in different representational systems: "Should we see such a [. . .] narrative process as one unit of meaning or two? Is it the equivalent of a sentence like 'The soldier shoots the villagers,' or of a formulation that expresses the soldier's agency less directly—for instance, 'The soldier fires. The villagers are shot'? But such an attempt at translating moving images into words cannot fully capture the difference. Filmic 'disconnection' has no parallel in language. It does have semiotic potential, however" (Ibid., p. 259).

The narration takes also the form of a progressively disclosing argumentation of the designer. When we act on an object on the screen, we are presented with a dynamic argument set forth by the designer (see Bogost, 2007). This argument is not presented on a single screen, as was the case of the visible language, but unfolds through multiple screens (globally). We call this type of narration a global narration. The global narration is supported by the navigation and interaction component running through a sequence of screens. These elements do not perceptually appear in some static form, but are enacted through the user's activities with the computer. The global narration can be further supported by design (and analysis) methods already known from theater (Laurel, 1993; Aristotle in Barnes, 1984; Barthes, 1977; Eco 1979), film (film grammar, as in Eisenstein, 1969) or comics (e.g., Cohn, 2005 and 2007 which applies a linguistic model to the sequence comics panels; see also McCloud, 1994).

The designer's argumentation can be supported also by different types of meta-communication, which further support the communication to the user by providing more details in the UI. The UI parts can be the accompanying documentation, help, tips, splash screens, introductions, *etc.* Designer's metacommunication, being a uni-directional communication (or a "one-shot-message" according to de Souza, 2005, p. 84), takes the form of narration, that augments what the screens by themselves say to the user.

The narrative perspective is used in different usability assessment methods for the user's part in the form of think-aloud protocols (e.g., de Souza, 2005) or scenarios (Carroll, 2000). The main benefit of the narration concept is that it directs the attention to the (textual, visual, etc.) flow of communication, which can be analyzed with proven semiotic and linguistic methods.

3.1.4 RHETORICAL TROPES

Rhetorical tropes are devices of persuasion and emphasis. As such they are employed by the HCI ideology. They are often present as metaphors. Several books about semiotics discuss rhetorical tropes, that is, figures of speech used for persuasion and emphasis, including one source that lists over 1000 tropes (Lanham, 1991). Marcus (1983) also discusses rhetorical tropes in visual communication.

Following on from the semiotics/linguistics framework, one can identify inter-action rhetorical tropes, or specific techniques of interaction and communication. These techniques include, among others, devices of substitution, namely metaphor, metonymy, prosopopoeia, and synecdoche. Metaphor "(a carrying across): some-thing is described as if it were something else" (Marcus, 1983, p. 4). Metonymy is "a type of metaphor in which an associated symbol is substituted the thing itself" (Ibid.). Prosopopoeia "is a personification of an inanimate object" (Ibid.). Synec-doche is a "substitution of a part for a whole or the whole for a part" (Ibid.). Their usefulness in interaction and communication design lies in the fact that they can lead the user to a "preferred reading" of the UI. This characteristic means the user can be directed to engage with the UI using a predefined set of interaction sentences.

The rhetorical tropes can also be revealed by intrinsic relations among the signs used, as is the case with "implied actions" (Apple, 2009). Implicit manipulation builds upon connotations of the desktop metaphor. An example would be the implicit relation between a document icon and the trash icon, or between a document icon and a printer icon. The implicit action is triggered by relating two objects by drag and drop. An object in a graphical UI is usually a document, folder, application, or trash can.

3.1.5 PATTERNS

Patterns are typical configurations of UI language components in different settings. The UI language components identified thus far, such as discrete elements, interactions sentences, rhetorical tropes, interaction games, narration, or interaction phases, can all form recurrent structures, that is, patterns. The number of the UI language elements should be as low as possible in order to minimize the cognitive load of the user. A useful guideline for the number of elements would be 7 ± 2, as presented by Miller (1956).

The signs on the just above-mentioned syntactic and lexical level all occur in certain relations (interdependency, determination, and constellation—Hjelmslev, 1961, cited from Andersen, 1997; see also Pimenta, 1997) more frequently than in others. This frequency stands at the root of pattern creation. Patterns can be further encapsulated within patterns. In such a way patterns can form superstructures, such as genres, or entire discourses.

The reciprocal relations among the UI signs follow certain rules contained in design guidelines. They can also form sets of UI patterns, some of which represent "best practices" (Erickson, 2000; Tidwell, 2006; Alexander, 1978; van Duyne et al., 2006). Such patterns (paradigms) bring about a certain context of use (or a complete scenario, see Carroll, 2000) and can therefore convey their meaning and usage with a greater complexity than purely syntactic elements. Therefore, these reciprocal relations can shift our attention, also, towards the syntactics/semantics plane.

As we shall see in the UI annotation and analysis chapter, interaction games form patterns that correspond to interaction genres. In our analysis, however, we shall first focus on the patterns based on scenarios with the help of a structured tool, the interaction sentence. Other types of patterns can be analyzed in future research.

3.2 UI ANNOTATION

In order to analyze the visible and interaction languages in a set of UIs from the interaction designer's and UI patterns perspective, it is important to extract the intrinsic set of rules (grammar). We assume, that taking advantage of the rules (of which language users may not be aware) to build UIs would lead to a more natural interaction flow and experience. All of the discrete elements should be taken into account (texts, pictures, etc.). See the multimodal approach used in Barthes' semiology analytical method (Barthes, 1977). For auditory icons or "earcons, " see Gaver (1986). For this purpose we shall use a formal notation (see Pimenta, 1997), which could be used together with the user–UI conversation transcript mentioned earlier.

Graphical (visible) notation is also an appealing way of visualizing interaction units. Among several researchers, we can mention Raj and Komaragiri (2009) (based on Alexander's patterns (Alexander, 1978) and the "inverted pyramid" by Cooper et al. (2007, p. 281); Cohn (2007; grammar of comics panels); or dance (or movement) grammar annotation by Asokan (2005) (see also movement notation "labanotation" by Rudolf Laban, in, e.g., Guest, 2005; for its application to computer interaction see, e.g., Loke et al., 2005).

However, for the present work we make use of a textual notation structured according to the interaction sentences mentioned in Section 3.1.2, "Interaction sentences." In the next step we could cross-compare several UIs and contrast their different approaches to designer's narration. Some would seem simple, we assume; others are overly complex or full of possibly useless redundancies. This method could also allow us to compare UIs from different periods of time and compare/contrast different epochs and their approach to UI storytelling/narration. These techniques could be used also to compare/contrast different cultures and their approach to UI narration: simple, direct, oriented to certain senses or manual, cognitive, emotional techniques.

As for the evaluation methods, we shall discuss the options available, as well as our choice in the following paragraphs. From the semiotic and UI language principles exposed so far we extracted a set of heuristics. We then developed a semiotic analysis (SA) method, that takes as input the interaction sentence transcript with figures from the UI. Because it is an evaluation method carried on by experts, we wanted to compare it with a well-known method to see, whether the results would be different, and how.

There are a number of expert-evaluation methods, including cognitive walkthrough, heuristic evaluation, expert inspection, and semiotic analysis. A comprehensive comparison of usability methods was done, for example, by Andre (2000), so we shall not go into much detail here. To compare the methods of expert evaluation we chose heuristic evaluation (HE) and semiotic analysis (SA). Our criteria for the methods were: fast and easy to do, results accessible to nonexperts, and comparable to previous data. The goal was to validate the SA against a nonsemiotic method. Although each of the methods follows a given set of heuristics, we expect them to have only few overlaps given their different emphasis.

Heuristic evaluation

The HEs considered are based on Marcus et al. (2003a); see Appendix A, "Heuristic Evaluation."

"Heuristic evaluation is a discount usability engineering method for quick, cheap, and easy evaluation of a user interface design. Heuristic evaluation is the most popular of the usability inspection methods. Heuristic evaluation is done as a systematic inspection of a user interface design for usability. The goal of heuristic evaluation is to find the usability problems in the design so that they can be attended to as part of an iterative design process. Heuristic evaluation involves having a small set of evaluators examine the interface and judge its compliance with recognized usability principles (the 'heuristics')" (Nielsen Norman Group, Heuristic Evaluation Articles and Training, `http://www.nngroup.com/topic/heuristic-evaluation`, cit. 2013-06-02).

The primary goal of an HE is to determine appropriately detailed failures or near-failures of usability (together with successes), measured informally against principles of usability, with direct citation of the principles involved, together with an informal grade of the severity of each.

We expect to gather the following evaluations and recommendations from the HE: significant errors, significant successes, recommended improvements, and prioritized actions.

Semiotic analysis

Semiotic analysis focuses on the signs present in the UI and extracts the underlying codes that make (or break) their meaning. A full specification of the method is included in Appendix B, "Semiotic Analysis."

The primary goal of the SA is "to establish the underlying conventions, identifying significant differences and oppositions in an attempt to model the system categories, relations (syntagmatic and paradigmatic), connotations, distinctions and rules of combination employed" (Chandler, 2001). We shall do this analysis by identifying the signs within the UI and the codes within which these signs have meaning, what sort of reality is produced, and what assumptions are made about the users. We shall also take into account rhetorical tropes.

For the assumptions about the user's part we can exploit the semiotic inspection, as proposed by de Souza (2006). In her research group (SERG: Semiotic Engineering Research Group) such inspection is carried out in five steps: "[1] an inspection of online and offline documentation and help content; [2] an inspection of static interface signs; [3] and inspection of dynamic interaction signs; [4] a contrastive comparison of designer-to-user metacommunications identified in steps [1], [2], [3]; and finally [5] a conclusive appreciation of the quality of the overall designer-to-user metacommunication" (de Souza, 2006, p. 149).

From the SA we expect to achieve more insights, capture more data, and gather more compatible data We expect the SA to be easier/faster to do; to be more amenable to cross-cultural use; to be more accessible/understandable to professional HCI/UX designers, usability analysts, business people, engineers, etc.; and to provide much previous data and analysis examples at a lower cost. For a discussion about the strengths of the semiotic method, see Chandler (2001).

The goal was to validate the SA against a nonsemiotic method. A validation between different semiotic methods can be made in a future study. As input for this comparison, we chose a UI corpus (Section 3.3) consisting of similar portions of two complex graphic design applications: Adobe Photoshop (PS) CS2 and the GNU Image Manipulation Program (GIMP) 2.6.7, both running on Mac OS X 10.5.8. The results gained both compare the usability of the applications during similar tasks, and the type outcome of both of the methods.

3.3 UI CORPUS

Because the chosen UIs are very complex, we need to select a sample body of material with which to work, or a *corpus* (see Barthes, 1977). There are various ways of selecting the corpus material. As Barthes suggests, "the corpus must be wide enough to give reasonable hope that its elements will saturate a complete system of resemblances and differences..." (Ibid., p. 97). Moreover, "the corpus must be as homogeneous as possible. To begin with, homogeneous in substance: there is an obvious interest in working on materials constituted by one and the same substance..." (Ibid.). "Further, homogeneous in time: in principle, the corpus must eliminate diachronic elements to the utmost..." (Ibid., p. 98).

As de Souza points out, the selection can follow different criteria. "One possibility is to choose portions that the design team thinks are most relevant, [...] that

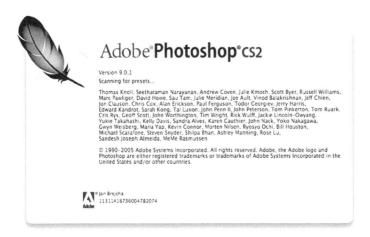

FIGURE 3.1 Adobe Photoshop splash screen. Source: Adobe Photoshop CS2. Adobe product screenshot(s) reprinted with permission from Adobe Systems Incorporated.

are commercially advertised [. . .], basic core functions which all users are likely to perform [. . .][or] a combination of criteria can also be used" (de Souza, 2006, p. 151).

As the UI corpus we selected the following:

Adobe Photoshop

For Adobe Photoshop CS2 (Adobe Systems, 2005) we can choose among the following product-description text (see Figure 3.1):

Vanishing Point.

Effortlessly edit and transform with tools that automatically adjust to the visual perspective of your images with the revolutionary, plane-based Vanishing Point. Define your perspective planes visually with the perspective grid tool, then paint, clone, and drag objects around corners and into the distance. Cutting literally hours off precision design and photo retouching tasks, one use and you'll wonder how you ever lived without it.

Spot Healing Brush.

Effortless, one-click retouching is yours with the advanced power of the new, Photoshop CS2 Spot Healing Brush. Click or paint flaws away, remove entire objects, and heal across all layers with your choice of blending and sample modes—even in 16-bit and CMYK[1] images. The advanced Spot Healing Brush analyzes the area around the tool as you use it, and automatically samples the best pixels to use for healing the clicked or painted area.

---

[1] CMYK refers to the four inks used in some color printing: cyan, magenta, yellow, and key (black).

One-Click Red Eye Tool.

Enjoy simple, accurate correction of the "red eye" caused by retinal flash reflection, with advanced control for pupil size and darkening effect when you want it.

Optical Lens Correction.

Quickly and efficiently correct a wide range of common camera lens distortion flaws from a single easy-to-use interface. Eliminate barrel or pincushion distortion, chromatic aberration, vignetting, and perspective flaws in all three dimensions, and do it in one pass with simple, intuitive controls, a live preview, and an alignment grid.

Multiple Layer Control with Smart Guides.

Intuitively select, align, group, and simultaneously move multiple layers on the document canvas. Drag a marquee around objects on multiple layers to create a multiple selection; shift-click to add new objects to the selection. Enable automatically-appearing Smart Guides to help line up your objects and snap them to borders and centers. Multiple Layer Control means enhanced efficiency.

Menu Customization.

Make it your Photoshop with customizable application menus. Photoshop CS2 comes equipped with many preset workspaces already optimized for specific task-based workflows, and you can quickly and easily create and save your own in the new Menu Customization window. Enable, disable, and color-key menu commands with your custom assigned keyboard shortcuts, then save your custom layout as a workspace for one-click access.

GIMP

For the GIMP (GNU Project, 2009) we can choose among the following product-description texts:

Customizable Interface.

Each task requires a different environment and GIMP allows you to customize the view and behavior the way you like it. Starting from the widget theme, allowing you to change colors, widget spacings, and icon sizes to custom tool sets in the toolbox. The interface is modularized into so-called docks, allowing you to stack them into tabs or keep them open in their own window. Pressing the tab key will toggle them hidden.

GIMP features a great full screen mode allowing you to not only preview your artwork but also do editing work while using the most of your screen estate (see also Figure 3.2).

Photo Enhancement.

Numerous digital photo imperfections can be easily compensated for using GIMP. Fix perspective distortion caused by lens tilt simply choosing the corrective mode in

FIGURE 3.2 GIMP splash screen. Source: The GIMP 2.6.7.

the transform tools. Eliminate lens' barrel distortion and vignetting with a powerful filter but a simple interface.

The included channel mixer gives you the flexibility and power to get your black-and-white photography to stand out the way you need.

Digital Retouching.
GIMP is ideal for advanced photo retouching techniques. Get rid of unneeded details using the clone tool, or touch up minor details easily with the new healing tool. With the perspective clone tool, it's not difficult to clone objects with perspective in mind just as easily as with the orthogonal clone.

Common features (PS CS2 and GIMP)

The following list of actions constitutes the semantics of the UI for the applications we selected. By aligning the semantic spaces of the application functions, we can better compare and analyze the other semiotic dimensions, such as syntactics and pragmatics. We aligned the features/tools (i.e., functions) by their title from the products' marketing publications and user manuals. The following list of actions constitutes the semantics of the selected UIs:

- Barrel distortion: Optical lens correction vs. Perspective clone
- Clone objects in perspective: Vanishing point vs. Perspective clone tool
- Customize the UI: Keyboard shortcuts and menu customization vs. Configure keyboard shortcuts
- Eliminate an object: Healing brush vs. Clone tool
- Reduce red eye: One-click red-eye tool vs. Red-eye removal

Note: we have not included in the "Top 5" list above the following functions:

- Colorize black and white: Shadow/highlight vs. Channel mixer
- Change resolution of file
- Move object to a layer
- Save image as PDF/PS
- Trim background

Following the above list, we are going to analyze each function by first transcribing the step-by-step process involved. Then, we shall focus on the HE and SA for each one. Because terminology differs considerably between the two products, in the transcript, we used only one version to make it concise.

4 Case Study: Expert Evaluations of Complex UIs

4.1 UI ANNOTATION AND ANALYSIS

We present our analysis results from the SA and HE of both of the compared UIs following the extracted UI corpus. We annotated the UIs using a transcript of interaction sentences from actions. The following subsections present our analysis of five selected functions from two image-processing applications.

4.1.1 RED-EYE REMOVAL

Adobe Photoshop steps

(0) Open the picture to adjust.
(1) Find the proper function in the menu bar or toolbox (Figure 4.1).

    (a) The subtask involved was to look through the menu items (especially in what seemed as most related: Image -> Adjustments and Filter) for a relative command, but it was not found.
    (b) The subtask involved was to look in the toolbox for a button resembling the intended action. Because many of the buttons with a similar function are grouped together (this fact is represented by a black triangle in the bottom-right corner of the button, however, only the first icon is shown as representing the group), the user needs to click and hold on every such group button to find out the group members.

(2) Click and hold the group button and select "Red Eye Tool" (Figure 4.2). (The cursor changes to a crosshair.)
(3) Click on the center of the red eye and repeat the same for other red eyes in the picture. (The eye color turns to a natural one.)
(4) Save the result in the file.

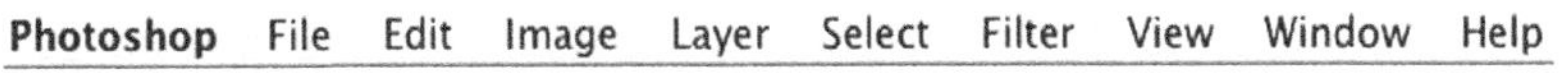

Photoshop   File   Edit   Image   Layer   Select   Filter   View   Window   Help

FIGURE 4.1 Adobe Photoshop menu bar. Source: Adobe Photoshop CS2. Adobe product screenshot(s) reprinted with permission from Adobe Systems Incorporated.

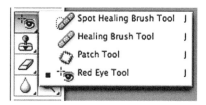

FIGURE 4.2   Adobe Photoshop red eye tool. Source: Adobe Photoshop CS2. Adobe product screenshot(s) reprinted with permission from Adobe Systems Incorporated.

HE analysis

Fitt's Law.

The buttons on the toolbar are quite small, even if they represent often-used functions. The toolbar layout supports rather advanced users familiar with accelerators.

Help and documentation.

The task 1b was not even described in the help documentation, so the user had to explore different possibilities by himself or herself.

Recognition rather than recall; perceived stability.

The button icon for the group comprising the "Red Eye Tool" changes every time a tool from that group is selected. Although the tools are semantically connected, their function and position on the menu has to be learned. This continual change of location could also affect the perceived stability of the UI.

SA analysis

Actors, audience, paradigm.

All of the objects involved in the interaction pertain to the leading paradigm of "window, icon, menu, pointing device" (WIMP). The paradigm is constituted by the menu bar, toolbars, main window containing the image, dialog windows, icons, and pointer. The paradigm is bound to the GUI metaphor. Adobe Photoshop is meant for professionals. This distinction of audience is manifested implicitly by the channel of distribution (commercial software) and explicitly in the marketing documentation (Adobe Photoshop's slogan reads: "The professional standard in desktop digital imaging" (Adobe, 2005)). The menu paradigm is constructed by combining noun-verb or verb-noun items, which seems deliberate (only one model should be chosen). A more specific audience for this function consists of photographers and home users.

Symbols.

The users are addressed by symbols pertaining to the user domain. In this case, the application icon and splash screen of Adobe Photoshop features a colorful feather. The connotations are elegance, simplicity, and naturalness which one would expect from a professional tool. What might break the expectation, however, is the historical usage of the image that symbolizes a writing pen. The other screens (and toolbars) are very compact and gray. The button activating the "Red Eye Tool" is found in the toolbox in the second group of tools from the top. All the tools in the group can modify the content of the picture in different ways (not cutting or cropping the picture, which modifies only the lexical dimension). The button is not readily visible, so the user has to click and hold the first icon of the subgroup, after which a list of all the included tools appears. The first subgroup icon has an arrow attribute in the bottom-right corner. The subgroup contains also two "retouching" brushes ("Healing Brush Tool" and "Spot Healing Brush Tool") and a patch tool.

Syntax.

The system processes are constituted by elements of interaction language. In the interaction transcript we can find many of the elements mentioned earlier. There are basic lexemes ("click," "option-click"), interaction sentences ("Open the picture to adjust"), rhetorical tropes (e.g., verb metaphors, such as "Red Eye Tool," symbolized by the combination of an indexical crosshair and symbolic human eye; metonymy, e.g., when clicking on the center of the eye applies the effect to the whole eye), and interaction games (these are the complete functions enabling us to achieve our goal, e.g., "Remove red eye"). The designer's narration element is found, for example, in the tool tip that helps reinforce an icon's meaning. Another example: In the status bar of the window or a dialogue window, text gives instruction about how to use the tool. In other dialogue windows, the UI presents the user with different choices, and finally, in the help menu, the UI describes the program functions comprehensively.

Rhetorical tropes.

Perhaps the most prominent of the rhetorical tropes in this context is the metaphor. The program metaphor builds upon the metaphor of a painter's canvas or photographer's studio. The product tries to transfer the environment into the present paradigm. Therefore, the image is placed on a "canvas," the pointer changes to different "brushes," the user can further apply different optical "filters," or use a choice of retouching "tools." By applying this approach, a number of inconsistencies emerge, which force users to twist or update their interpretation of the metaphor. The canvas, for example, is in fact infinite and can be resized in different ways at any time. The picture "lying" on the canvas can consist of infinite layers. Almost any tool can be customized using the "brush" metaphor: One can modify the thickness, shape, or profile of the brush. A filter can be used afterwards, applied as a part of retouching. More fundamentally, time can be manipulated also through the "undo" function that steps back through the history of actions.

Interaction phases.

On the level of interaction sentence, the interaction changes to reflect the constant evaluation of results on the user's part. The interaction sentence is then modified or repeated accordingly. Considering the example from the transcripts, the action is modified after the system's feedback (when clicking on the plane to clone with the clone tool, the user is instructed to option-click on the source plane first), the action is repeated (click on several instances of red eyes in the picture), or the action is needed only once (when applying the changes by pressing the "OK" button). The middle of

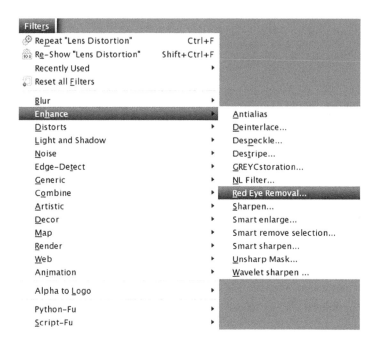

FIGURE 4.3 GIMP red eye tool. Source: The GIMP 2.6.7.

the interaction game differs from the beginning and end because a new window is shown, keeping the user accordingly away from the picture he or she opened.

GIMP steps

(0) Open the picture to adjust.

(1) Find the proper function in the menu bar or toolbar.

    (a) The subtask involved was to look at the toolbar for a button resembling the intended action. No such button was found. The buttons were presented on a single layer (not in groups as in PS CS2).

    (b) Then look through the menu items (especially in what seemed as most related: Image, and Filters -> Enhance) for a relative command, here it was "Red Eye Removal..." (see Figures 4.3 and 4.4). (A dialog box appeared.)

(2) Adjust the threshold slider and click "OK." There's a live preview showing how the image will be affected after clicking "OK." The dialog also says in a tip, that "Manually selecting the eyes may improve the results."

(3) Save the results in the file.

Alternatively the user can (2a) select the red eye using a selection tool (lasso or elliptical), (2b) invoke the red eye removal dialog, (2c) set the threshold needed, and click "OK." This sequence should be repeated for other red eyes in the picture.

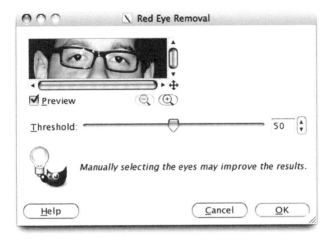

FIGURE 4.4 GIMP red eye removal. Source: The GIMP 2.6.7.

HE analysis

Recognition rather than recall.

The dialog in (2) does not provide the information about how to select the eyes. Clicking on the eyes from within the dialog does not affect the image preview. Only in the Help is there a further explanation saying the user "must do a selection (lasso or elliptical) of the boundary of the iris of the eye(s) having a red pupil."

Modelessness.

A window to remove the red eyes is shown; the window locks and constrains the user into a mode where he or she has to select some parameters in order to continue working on the image. The dialog window can be even invoked more than once, which makes it hard to understand whether the proposed changes in the previews in different windows will be applied simultaneously or on top of each other.

SA analysis

Actors, audience, paradigm.

All of the objects involved in the interaction pertain to the leading paradigm of "window, icon, menu, pointing device" (WIMP). The paradigm is constituted by the menu bar, toolbars, main window containing the image, dialog windows, icons, and pointer. The paradigm is bound to the GUI metaphor. GIMP is intended for amateurs/semiprofessionals and programmers. This distinction is manifested implicitly by the channel of distribution (open-source) and explicitly in the marketing documentation (in GIMP it is by stating, that "[i]n the free software world, there is generally no distinction between users and developers" (GNU Project, 2009)). The menu paradigm is constructed by combining noun-verb or verb-noun items, which seems deliberate (only one model should be chosen). GIMP shares the same user group as Adobe Photoshop (photographers and advertising designers).

Symbols.

The users are addressed by symbols related to the user domain. In this case, GIMP's icon of a stylized dog head connotes playfulness, fun, and also ease of use. The icon is not used on the splash screen, however, in favor of a planet picture. The toolbars and other screens show larger and more colorful button icons and larger dialog windows, which are easily reached by the pointer. The icon symbols used in the menus, for example, in the "Tools" menu, make no distinction between nouns (e.g., Pencil, Eraser, Text) and verbs (e.g., Zoom, Measure, Heal) which could be helpful. Also, the symbols are created by different methods (e.g., the Pencil tool has an iconic representation of a pen but the Zoom tool icon is created by metonymy with its action and uses a zooming lens; other are connected only loosely, as in the case of Swap Colors).

Syntax.

The system processes are constituted by the same UI language components analyzed above for Adobe Photoshop. In GIMP, there is only a difference in the tool metaphor used ("Remove red eyes"). The designer's narration element is found in the dialogue window, where it gives instruction about how to use the tool ("Manually selecting the eyes may improve the results") and presents the user with different choices.

Rhetorical tropes.

Perhaps the most prominent of the rhetorical tropes in this context is the metaphor. As is the case of syntax, the same set of metaphors is shared with Adobe Photoshop.

Interaction phases.

The interaction phases are similar to those mentioned above in the Adobe Photoshop analysis. Also, the interaction sentence level is similar. Taking the example from the transcripts, the action is modified after the system's feedback (when selecting the tool the user is instructed, that "Manually selecting the eyes may improve the results," the action is repeated (apply the tool on every red eye separately), or the action was needed only once (when letting the system apply the filter automatically and clicking the "OK" button). However, to get the best results, the user is forced to return to his or her previous state and select the eyes manually first, thus traversing the natural timeline.

4.1.2 BARREL DISTORTION

Adobe Photoshop steps

(0) Open the picture to adjust.
(1) Find the proper function in the menu bar or toolbox.

   (a) The subtask involved was to look at the toolbox for a button resembling the intended action. Nothing like that was found.

(b) Alternatively look through the menu items (especially in what seemed as most related: Image -> Adjustments, and Filter -> Distort) for a relative command (it was found under Filter -> Distort -> Lens correction). (A window with a live preview, "Remove distortion" sliders, and "Transform" sliders appears.)

(2) Drag the sliders to set the needed parameters.
(3) Click "OK" to apply the changes.
(4) From the menu select "Filter -> Distort -> Shear" for an additional correction. (A window "Shear" appears with a draggable curve, and a live preview.)
(5) Drag the curve to provide the intended distortion. See the proposed results in the preview.
(6) Click "OK" to apply the changes.
(7) Save changes to the file.

HE analysis

Visible interfaces/WYSIWYG.

The lens correction function was not present on the toolbar, and was only accessible through the menu bar. Since it is one of the advertised features, it should be accessible as readily as possible.

Direct manipulation/see and point.

To make any changes to the image perspective the user can't interact with the image directly. Instead, he or she is presented with a window containing a set of controllers. All the actions are only visible in the preview window.

Modelessness.

By presenting the user with a "lens correction" window the user is not able to interact freely with the image. In order to continue working he or she must dismiss the dialog window first (see Figures 4.5 and 4.6). A better solution seems to be using standard controls and not introducing a different working environment. By doing so, we would also eliminate the extra step of applying the changes.

SA analysis

Actors, audience, paradigm.

All of the objects involved in the interaction pertain to the leading paradigm of "window, icon, menu, pointing device" (WIMP). The paradigm is constituted by the menu bar, toolbars, main window containing the image, dialog windows, icons, and pointer. The paradigm is bound to the GUI metaphor. Adobe Photoshop is meant for professionals. This distinction of audience is manifested implicitly by the channel of distribution (commercial software) and explicitly in the marketing documentation (Adobe Photoshop's slogan reads: "The professional standard in desktop digital imaging" (Adobe, 2005)). The menu paradigm is constructed by combining noun-verb or verb-noun items, which seems deliberate (only one model should be chosen.)

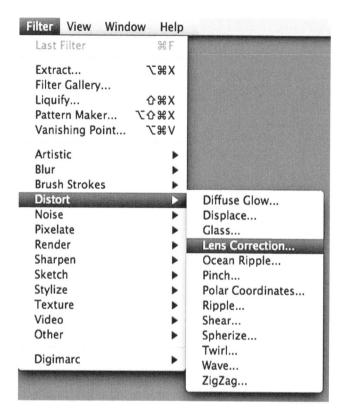

FIGURE 4.5 Adobe Photoshop lens correction. Source: Adobe Photoshop CS2. Adobe product screenshot(s) reprinted with permission from Adobe Systems Incorporated.

A more specific audience for this function consists of photographers and advertising designers.

Symbols.

The users are addressed by symbols pertaining to the user domain. In this case, the application icon and splash screen of Adobe Photoshop features a colorful feather. The connotations are elegance, simplicity, and naturalness, which one would expect from a professional tool. What might break the expectation, however, is the historical usage of the image that symbolizes a writing pen. The other screens (and toolbars) are very compact and gray. The menus are only text-based, whereas the toolbar has only icons (with a textual label). The icons in the toolbar are related to their object in different ways but are connected to the prevailing metaphor and follow the application genre conventions.

Syntax.

The system processes are constituted by UI language components, as described earlier. In the interaction transcript, we can find all the elements mentioned. There are basic lexemes ("click," "option-click"), interaction sentences ("Open the picture to adjust"),

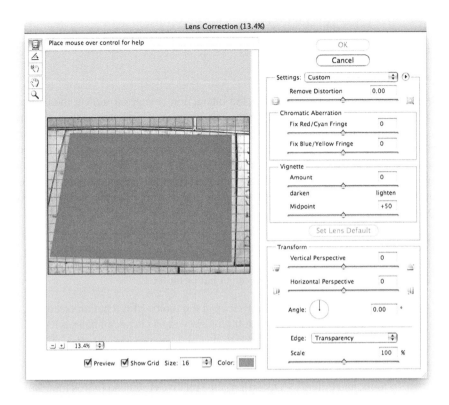

FIGURE 4.6 Adobe Photoshop lens correction window. Source: Adobe Photoshop CS2. Adobe product screenshot(s) reprinted with permission from Adobe Systems Incorporated.

rhetorical tropes (e.g., in Synecdoche, such as Hand Tool for panning the picture in the preview), interaction games (these are the complete functions enabling us to accomplish our goal, e.g., "Filter -> Distort -> Lens correction. . ."). The designer's narration element is found in the dialogue window when it presents the user with different choices, and finally, in the Help menu it describes the program functions comprehensively.

Rhetorical tropes.

Perhaps the most prominent of the rhetorical tropes in this context is the metaphor. The program metaphor builds upon the metaphor of a painter's canvas or photographer's studio. The product tries to transfer the environment into the present paradigm. Therefore, the image is placed on a "canvas," the pointer changes to different "brushes," the user can further apply different optical "filters," or use a choice of retouching "tools." By applying this approach, a number of inconsistencies emerge, which force users to twist or update their interpretation of the metaphor. The canvas, for example, is in fact infinite and can be resized in different ways at any time. The picture "lying" on the canvas can consist of infinite layers. Almost any tool can be customized using the "brush" metaphor. One can modify the thickness, shape, or profile of the brush.

A filter can be used afterwards, applied as a part of retouching. More fundamentally, time can be manipulated also through the "undo" function that steps back through the history of actions.

Interaction phases.

On the level of interaction sentence, the interaction changes to reflect the constant evaluation of results on the user's part. The interaction sentence is then modified or repeated accordingly. Considering the example from the transcripts, the action is modified after the system's feedback (when the system shows in the preview window the proposed change), or the action was needed only once (when clicking the "OK" button). The middle of the interaction game differs from the beginning and end because a new window is shown keeping the user accordingly away from the picture he or she opened.

GIMP steps

(0) Open the picture to adjust.
(1) Find the proper function in the menu bar or toolbox.

    (a) The subtask involved was to look at the toolbox for a button resembling the intended action. A button showing a change of perspective was found.

    (b) Alternatively look through the menu items (especially in what seemed as most related: Image -> Adjustments, Tools -> Transformation tools, and Filters -> Distortion) for a relative command (it was found under Tools -> Transformation tools -> Perspective). (A four-button floating window has opened, the pointer changed, and the image acquired boxes on each corner.)

(2) Drag the boxed corner of the image in the direction that would compensate the perspective. Repeat this for every corner, as necessary. (The live preview shows the proposed transformation).
(3) Click "Transform" in the floating window "Perspective." (The transformation is applied). (See Figures 4.16 and 4.8.)
(4) Select "Filters -> Distortion -> Lens distortion" to make further needed corrections. (A window with a live preview and six sliders appear.)
(5) Drag each of the sliders to set the needed parameter. See the proposed result in the preview.
(6) Click "OK" to apply the filter.
(7) Save changes to the file.

HE analysis

Direct manipulation/See and point.

Although the perspective of the image could be changed by dragging the mouse, making further needed changes to the image perspective was only possible through a window containing a set of controllers. All the actions are only visible in the preview window.

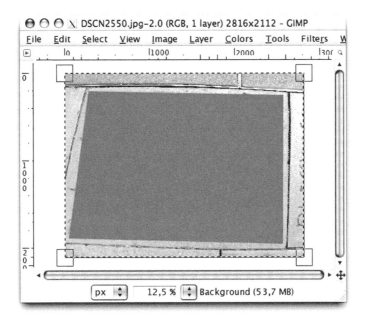

FIGURE 4.7 GIMP perspective function window with affordance boxes. Source: The GIMP 2.6.7.

Modelessness.

By presenting the user with a "lens correction" window the user is not able to interact freely with the image. In order to continue working he or she must dismiss the window first.

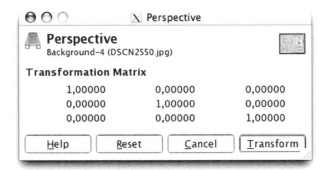

FIGURE 4.8 GIMP perspective transformation window. Source: The GIMP 2.6.7.

SA analysis

Actors, audience, paradigm.

All of the objects involved in the interaction pertain to the leading paradigm of "window, icon, menu, pointing device" (WIMP). The paradigm is constituted by the menu bar, toolbars, main window containing the image, dialog windows, icons, and pointer. The paradigm is bound to the GUI metaphor. GIMP is intended for amateur/semiprofessionals and programmers. This distinction is manifested implicitly by the channel of distribution (open-source) and explicitly in the marketing documentation (in GIMP it is by stating, that "[i]n the free software world, there is generally no distinction between users and developers" (GNU Project, 2009)). The menu paradigm is constructed by combining noun-verb or verb-noun items, which seems deliberate (only one model should be chosen). GIMP shares the same user group as Adobe Photoshop (photographers and advertising designers).

Symbols.

The users are addressed by symbols related to the user domain. In this case, GIMP's icon of a stylized dog head connotes playfulness, fun, and ease of use. The icon is not used on the splash screen, however, in favor of a planet picture. The toolbars and other screens show larger and more colorful button icons and larger dialog windows that are easily reached by the pointer. The icon symbols used in the menus, e.g., in the "Tools" menu, make no distinction between nouns (e.g., Pencil, Eraser, Text) and verbs (e.g., Zoom, Measure, Heal), which could be helpful. Also, the symbols are created by different methods (e.g., the Pencil tool has an iconic representation of a pen but the Zoom tool icon is created by metonymy with its action and uses a zooming lens; others are connected only loosely, as in the case of Swap Colors).

Syntax.

The system processes are constituted by the same UI language components analyzed above for Adobe Photoshop. In GIMP, there is only a difference in the rhetorical tropes (e.g., visual explanation, when the "perspective tool" icon shows a square pulled in the direction of arrows), and interaction games (these are the complete functions enabling us to accomplish our goal, e.g., "correct image perspective").

Rhetorical tropes.

Perhaps the most prominent of the rhetorical tropes in this context is the metaphor. As is the case of syntax, the same set of metaphors is shared with Adobe Photoshop.

Interaction phases.

The interaction phases are similar to those mentioned above in the Adobe Photoshop analysis. Also, the interaction sentence level is similar. However, the middle phase (where the user works on the picture) seems to be more consistent with beginning and end. This is because the user keeps working in the image window and is not distracted by other windows or palettes. Considering the example from the transcripts, the action is modified after the system's feedback (when adjusting the perspective the system

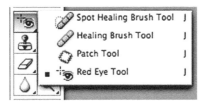

FIGURE 4.9 Adobe Photoshop healing brush. Source: Adobe Photoshop CS2. Adobe product screenshot(s) reprinted with permission from Adobe Systems Incorporated.

recalculates the change and shows it on the canvas before it can be applied), the action is repeated (drag the tool several times in different directions until the perspective is adjusted), or the action was needed only once (when clicking the "OK" button).

4.1.3 ELIMINATE AN OBJECT

Adobe Photoshop steps

(0) Open the picture to adjust.
(1) Find the proper function in the menu bar or toolbox.

    (a) The subtask involved was to look at the toolbox for a button resembling the intended action. It was found in the group containing "Red Eye Tool" as "Healing Brush Tool." (See Figures 4.9–4.11.)

    (b) Alternatively look through the menu bar items (especially in what seemed as most related: Image -> Adjustments, and Filter) for a relative command, but nothing was found. (An options bar appears with the parameters of the brush.)

(2) Click on the object to eliminate. (A dialog window appears saying to option-click to define the source.)
(3) Click "OK."
(4) Option-click on the source.

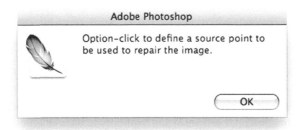

FIGURE 4.10 Adobe Photoshop healing dialog. Source: Adobe Photoshop CS2. Adobe product screenshot(s) reprinted with permission from Adobe Systems Incorporated.

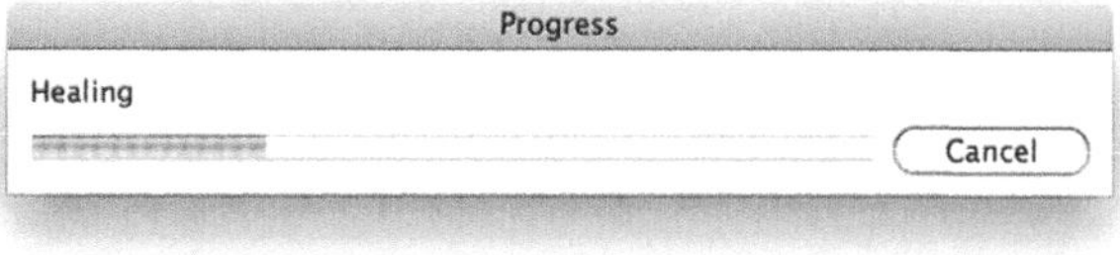

FIGURE 4.11 Adobe Photoshop healing process. Source: Adobe Photoshop CS2. Adobe product screenshot(s) reprinted with permission from Adobe Systems Incorporated.

(5) Drag the pointer (brush) over the object to eliminate. Repeat several times until the object is canceled. (Every time the mouse button is released a dialog appears with the healing progress bar.)

(6) Save changes to the file.

HE analysis

Direct manipulation/see and point; error prevention.

Although the users can use the tool directly on the image, they are reminded every time to select a source region first. Instead of forcing the user to go "backwards," the program should allow the user to select the region afterwards. Such a change in the perceived interaction timeline also violates the principle of error prevention.

Recognition rather than recall.

The healing function works to remove texture imperfections of the image. The user, however, is not presented with any clue that he or she can use it also to completely remove an object.

Match between system and the real world.

The "plaster" symbol in the real world works to cover a wound and help it to heal. In the system it is not covering nor healing anything. It is only modifying the pixel texture.

SA analysis

Actors, audience, paradigm.

All of the objects involved in the interaction pertain to the leading paradigm of "window, icon, menu, pointing device" (WIMP). The paradigm is constituted by the menu bar, toolbars, main window containing the image, dialog windows, icons, and pointer. The paradigm is bound to the GUI metaphor. Adobe Photoshop is meant for professionals. This distinction of audience is manifested implicitly by the channel of distribution (commercial software) and explicitly in the marketing documentation (Adobe Photoshop's slogan reads: "The professional standard in desktop digital imaging" (Adobe, 2005)). The menu paradigm is constructed by combining noun-verb or verb-noun items, which seems deliberate (only one model should be chosen.) A more specific audience for this function is photographers and advertising designers.

Symbols.

The users are addressed by symbols pertaining to the user domain. In this case, the application icon and splash screen of Adobe Photoshop feature a colorful feather. The connotations are elegance, simplicity, and naturalness, which one would expect from a professional tool. What might break the expectation, however, is the historical usage of the image that symbolizes a writing pen. The other screens (and toolbars) are very compact and gray. The menus are only text-based, whereas the toolbar has only icons (with a textual label). The icons in the toolbar are related to their object in different ways but are connected to the prevailing metaphor and follow the application genre conventions.

Syntax.

The system processes are constituted by UI language components, as described earlier. In the interaction transcript, we can find all the elements mentioned. There are basic lexemes ("click," "option-click"), interaction sentences ("Open the picture to adjust"), rhetorical tropes (e.g., metaphors, such as "Healing brush"), and interaction games (these are the complete functions enabling us to accomplish our goal, e.g., "Eliminate an object"). The designer's narration element is found in the tool-tip help reinforcing the icon meaning, in the status bar of the window or a dialog window for which help is given by instructions regarding use of the tool and in other dialog windows, which presents the user with different choices. Finally, in the Help menu, the text comprehensively describes the program functions.

Rhetorical tropes.

Perhaps the most prominent of the rhetorical tropes in this context is the metaphor. The program metaphor builds upon the metaphor of a painter's canvas or photographer's studio. The product tries to transfer the environment into the present paradigm. Therefore, the image is placed on a "canvas," the pointer changes to different "brushes," the user can further apply different optical "filters," or use a choice of retouching "tools." By applying this approach, a number of inconsistencies emerge, which force users to twist or update their interpretation of the metaphor. The canvas, for example, is in fact infinite and can be resized in different ways at any time. The picture "lying" on the canvas can consist of infinite layers. Almost any tool can be customized using the "brush" metaphor: one can modify the thickness, shape, or profile of the brush. A filter can be used afterwards, applied as a part of retouching. More fundamentally, time can be manipulated also through the "undo" function that steps back through the history of actions.

Interaction phases.

On the level of interaction sentence, the interaction changes to reflect the constant evaluation of results on the user's part. The interaction sentence is then modified or repeated accordingly. Considering the example from the transcripts, the action is modified after the system's feedback (when clicking on the object to eliminate with the healing tool the user is instructed to option-click on the source first), the action is

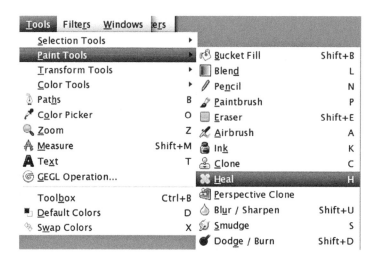

FIGURE 4.12 GIMP heal tool. Source: The GIMP 2.6.7.

repeated (drag the brush several times over the object to eliminate until the object is canceled), or the action was needed only once (when clicking the "OK" button).

GIMP steps

(0) Open the picture to adjust.

(1) Find the proper function in the menu bar or toolbox.

    (a) The subtask involved was to look at the toolbox for a button resembling the intended action. It was found as healing brush with a symbol of a "healing plaster." (See Figures 4.12 and 4.13.)

    (b) Alternatively look through the menu bar items (especially in what seemed as most related: Image -> Adjustments, Filters, Tools) for a relative command. It was found under Tools -> Drawing tools -> Healing. (The pointer changed and gained an attribute showing the selected tool.)

(2) Click on the object to work on. (A warning in the status bar appears telling to select a source first using control-click.)

(3) Control-click on the source.

(4) Drag the pointer (brush) over the object to eliminate. (Every time the mouse button is released the effect is applied. This tool, however, is intended just for small corrections, as we learn from the manual, so applying another tool before is it necessary.)

(5) Undo the changes (and use another tool).

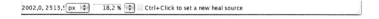

FIGURE 4.13 GIMP heal status bar. Source: The GIMP 2.6.7.

(6) Select the Clone tool (symbolized by a stamp) from the toolbox.

(7) Drag the pointer (brush) over the object to eliminate. Repeat several times until the object is canceled. (Every time the mouse button is released the effect is applied.)

(8) Repeat with the healing brush afterwards.

(9) Save changes to the file.

HE analysis

Direct manipulation/see and point; error prevention.

Although the user can use the tool directly on the image, he or she is reminded every time to select a source region first. Instead of forcing the user to go "backwards," the program should allow the user to select the region afterwards. Such a change in the perceived interaction timeline also violates the principle of error prevention.

Recognition rather than recall.

The healing function works of course to remove texture imperfections of the image. The user, however, is not presented with any clue that he or she can use it also to completely remove an object.

Match between system and the real world.

The "plaster" symbol in the real world works to cover a wound and help it to heal. In the system it is not covering nor healing anything. It is only modifying the pixel texture. Moreover, the interface didn't provide information to use after the clone tool.

SA analysis

Actors, audience, paradigm.

All of the objects involved in the interaction pertain to the leading paradigm of "window, icon, menu, pointing device" (WIMP). The paradigm is constituted by the menu bar, toolbars, main window containing the image, dialog windows, icons, and pointer. The paradigm is bound to the GUI metaphor. GIMP is intended for amateur/semiprofessionals and programmers. This distinction is manifested implicitly by the channel of distribution (open-source) and explicitly in the marketing documentation (in GIMP it is by stating, that "[i]n the free software world, there is generally no distinction between users and developers" (GNU Project, 2009)). The menu paradigm is constructed by combining noun-verb or verb-noun items, which seems deliberate (only one model should be chosen). GIMP shares the same user group as Adobe Photoshop (photographers and advertising designers).

Symbols.

The users are addressed by symbols related to the user domain. In this case, GIMP's icon of a stylized dog head connotes playfulness, fun, and ease of use. The icon is not used on the splash screen, however, in favor of a planet picture. The toolbars and other screens show larger and more colorful button icons and larger dialog windows that are easily reached by the pointer. The icon symbols used in the menus, e.g., in the "Tools" menu, make no distinction between nouns (e.g., Pencil, Eraser, Text) and verbs (e.g., Zoom, Measure, Heal), which could be helpful. Also, the symbols are created by different methods (e.g., the Pencil tool has an iconic representation of a pen but the Zoom tool icon is created by metonymy with its action and uses a zooming lens; others are connected only loosely, as in the case of Swap Colors).

Syntax.

The system processes are constituted by the same UI language components analyzed above for Adobe Photoshop.

Rhetorical tropes.

Perhaps the most prominent of the rhetorical tropes in this context is the metaphor. As is the case of syntax, the same set of metaphors is shared with Adobe Photoshop.

Interaction phases.

The interaction phases are similar to those mentioned above in the Adobe Photoshop analysis. Also, the interaction sentence level is similar. Considering the example from the transcripts, the action is modified after the system's feedback (when clicking on the object to eliminate with the healing tool the user is instructed to option-click on the source first), the action is repeated (drag the brush several times over the object to eliminate until the object is canceled), or the action was needed only once (when clicking the "OK" button).

4.1.4 CLONE OBJECTS IN PERSPECTIVE

Adobe Photoshop steps

(0) Open the picture to adjust.

(1) Find the proper function in the menu or tool palette.

    (a) The subtask involved was to look at the toolbox for a button resembling the intended action, but it was not found.

    (b) Alternatively look through the menu items (especially in what seemed as most related: Image -> Adjustments, and Filter) for a relative command (it was found under Filter -> Vanishing point. . .). (A window called "Vanishing Point" appears. The window sports a live preview, "Create Plane Tool," and "Clone Tool," among others.) (See Figures 4.14 and 4.15.)

(2) Click the four corners according to the information line provided ("Click the four corners of a perspective plane or object in the image to create an editing plane. Tear off perpendicular planes from the stretch nodes of existing. . .)."

(3) Select the "Clone tool."

(4) Option-click in the plane to set the source. ("Opt+click in a plane, to set a source point for the clone. Once source point is set, click+drag to paint or clone. Shift+click to extend the stroke to last click".)

FIGURE 4.14 Adobe Photoshop vanishing point menu. Source: Adobe Photoshop CS2. Adobe product screenshot(s) reprinted with permission from Adobe Systems Incorporated.

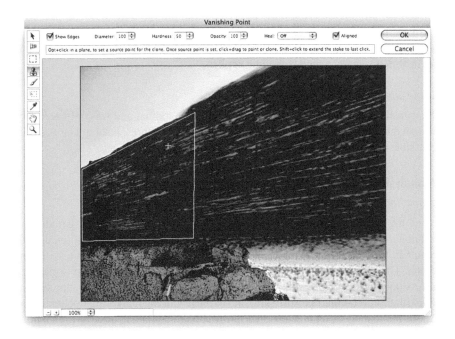

FIGURE 4.15 Adobe Photoshop vanishing point result. Source: Adobe Photoshop CS2. Adobe product screenshot(s) reprinted with permission from Adobe Systems Incorporated.

(5) Click-drag (to paint) several times to clone in the perspective. See the proposed results in the preview.
(6) Click "OK" to apply the changes.
(7) Save changes to the file.

HE analysis

Direct manipulation/see and point; error prevention.

Although the users can use the tool directly on the image, they are reminded every time to select a source region first. Instead of forcing the user to go "backwards," the program should allow the user to select the region afterwards. Such a change in the perceived interaction timeline also violates the principle of error prevention.

Modelessness.

By selecting the vanishing point function, the user is presented with a new window (named "Vanishing Point") containing the image to manipulate and a reduced set of controls (buttons, check-boxes, and drop-down menus). After the adjustments, the user has to click "OK" to transfer the changes to the image in the main window underneath. A better solution seems to be using standard controls and not introducing a different working environment. By doing so, we would also eliminate the extra step of applying the changes.

Recognition rather than recall.

All of the needed actions are visible and the system provides inline help. However, the toolbar on the top-left does not show which tools are necessary for the operation and in which sequence they should be applied.

Visible interfaces/WYSIWYG.

The vanishing point function was not present on the toolbar and was only accessible through the menu bar. Since it is one of the advertised features, it should be as readily accessible as possible.

SA analysis

Actors, audience, paradigm.

All of the objects involved in the interaction pertain to the leading paradigm of "window, icon, menu, pointing device" (WIMP). The paradigm is constituted by the menu bar, toolbars, main window containing the image, dialog windows, icons, and pointer. The paradigm is bound to the GUI metaphor. Adobe Photoshop is meant for professionals. This distinction of audience is manifested implicitly by the channel of distribution (commercial software) and explicitly in the marketing documentation (Adobe Photoshop's slogan reads: "The professional standard in desktop digital imaging" (Adobe, 2005)). The menu paradigm is constructed by combining noun-verb or verb-noun items, which seems deliberate (only one model should be chosen.) A more specific audience for this function are photographers and advertising designers.

Symbols.

The users are addressed by symbols pertaining to the user domain. In this case, the application icon and splash screen of Adobe Photoshop feature a colorful feather. The connotations are elegance, simplicity, and naturalness, which one would expect from a professional tool. What might break the expectation, however, is the historical usage of the image that symbolizes a writing pen. The other screens (and toolbars) are very compact and gray. The menus are only text-based, whereas the toolbar has only icons (with a textual label). The icons in the toolbar are related to their object in different ways but are connected to the prevailing metaphor and follow the application genre conventions.

Syntax.

The system processes are constituted by UI language components, as described earlier. In the interaction transcript, we can find all the elements mentioned. There are basic lexemes ("click," "option-click"), interaction sentences ("Open the picture to adjust"), rhetorical tropes (e.g., metaphors, such as "Vanishing Point" or "Clone Tool"), interaction games (these are the complete functions enabling us to accomplish our goal, e.g., "Clone an object in perspective"). The designer's narration element is found in the tool-tip help reinforcing the icon meaning, in the status bar of the window, or a dialog window for which help is given by instructions regarding use of the tool and in other dialog windows that present the user with different choices. Finally, in the Help menu, the text comprehensively describes the program functions. In the "Vanishing Point" window, the designer's narration gives detailed instructions for all the steps involved.

Rhetorical tropes.

Perhaps the most prominent of the rhetorical tropes in this context is the metaphor. The program metaphor builds upon the metaphor of a painter's canvas or photographer's studio. The product tries to transfer the environment into the present paradigm. Therefore, the image is placed on a "canvas," the pointer changes to different "brushes," the user can further apply different optical "filters," or use a choice of retouching "tools." By applying this approach, a number of inconsistencies emerge, which force users to twist or update their interpretation of the metaphor. The canvas, for example, is in fact infinite and can be resized in different ways at any time. The picture "lying" on the canvas can consist of infinite layers. Almost any tool can be customized using the "brush" metaphor: One can modify the thickness, shape, or profile of the brush. A filter can be used afterwards, applied as a part of retouching. More fundamentally, time can be manipulated also through the "undo" function that steps back through the history of actions.

Interaction phases.

On the level of interaction sentence, the interaction changes to reflect the constant evaluation of results on the user's part. The interaction sentence is then modified or repeated accordingly. Considering the example from the transcripts, the action is modified after the system's feedback (when clicking on the plane to clone with the clone tool, the user is instructed to option-click on the source plane first), the action is

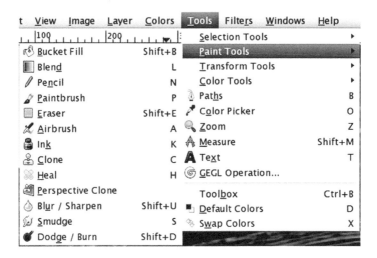

FIGURE 4.16 GIMP perspective clone menu. Source: The GIMP 2.6.7.

repeated (drag the brush several times to paint the object in the new perspective), or the action is needed only once (when applying the changes by pressing the "OK" button). The middle of the interaction game differs from the beginning and end because a new window is shown, keeping the user accordingly away from the picture he or she opened.

GIMP steps

(0) Open the picture to adjust.

(1) Find the proper function at the menu or tool palette.

    (a) The subtask involved was to look through the menu items (especially in what seemed as most related: Image -> Adjustments, Filters, and Tools -> Transform Tools) for a relative command (it was found under Tools -> Paint Tools -> Perspective Clone).

    (b) Alternatively look in the toolbox for a button resembling the intended action. It was found as Perspective Clone. (Boxes on each corner of the image appeared, the pointer changed to crosshair with the tool attribute, and the toolbox expanded to show the "Modify Perspective" selected and the "Perspective Clone" radio button.) (See Figures 4.16 and 4.17.)

(2) Drag the four corner boxes to define the perspective plane to clone.

(3) Click on the "Perspective Clone" radio button to change the tool.

(4) Control-click the source in the defined plane.

(5) Click-drag (to paint) several times to clone in the perspective.

(6) Save changes to the file.

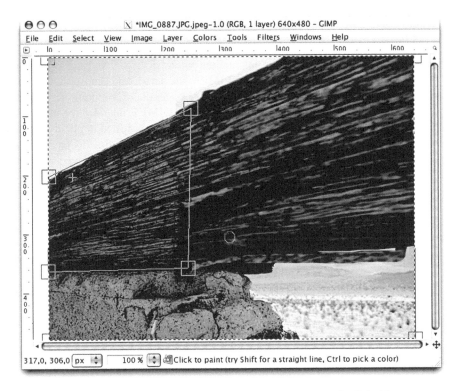

FIGURE 4.17 GIMP perspective clone plane and tool. Source: The GIMP 2.6.7.

HE analysis

Direct manipulation/see and point; error prevention.

Although the user can use the tool directly on the image, he or she is reminded every time to select a source region first. Instead of forcing the user to go "backwards," the program should allow the user to select the region afterwards. Such a change in the perceived interaction timeline also violates the principle of error prevention.

Recognition rather than recall.

All of the needed actions are visible and the system provides inline help. However, the toolbar on the top-left does not show which tools are necessary for the operation and in which sequence they should be applied ("Modify Perspective" or "Perspective Clone"?).

Match between system and real world.

The Perspective Clone tool is located under Paint Tools and thus supports the metaphor of painting on the picture. However, in the virtual environment this could be problematic as the clone tool is connected with image transformation and/or filtering.

SA analysis

Actors, audience, paradigm.

All of the objects involved in the interaction pertain to the leading paradigm of "window, icon, menu, pointing device" (WIMP). The paradigm is constituted by the menu bar, toolbars, main window containing the image, dialog windows, icons, and pointer. The paradigm is bound to the GUI metaphor. GIMP is intended for amateurs, semiprofessionals and programmers. This distinction is manifested implicitly by the channel of distribution (open-source) and explicitly in the marketing documentation (in GIMP it is by stating, that "[i]n the free software world, there is generally no distinction between users and developers" (GNU Project, 2009)). The menu paradigm is constructed by combining noun-verb or verb-noun items, which seems deliberate (only one model should be chosen). GIMP shares the same user group as Adobe Photoshop (photographers and advertising designers).

Symbols.

The users are addressed by symbols related to the user domain. In this case, GIMP's icon of a stylized dog head connotes playfulness, fun, and ease of use. The icon is not used on the splash screen, however, in favor of a planet picture. The toolbars and other screens show larger and more colorful button icons and larger dialog windows that are easily reached by the pointer. The icon symbols used in the menus, e.g., in the "Tools" menu, make no distinction between nouns (e.g., Pencil, Eraser, Text) and verbs (e.g., Zoom, Measure, Heal), which could be helpful. Also, the symbols are created by different methods (e.g., the Pencil tool has an iconic representation of a pen but the Zoom tool icon is created by metonymy with its action and uses a zooming lens; other are connected only loosely, as in the case of Swap Colors).

Syntax.

The system processes are constituted by the same UI language components analyzed above for Adobe Photoshop. In GIMP, there is only a difference in the tool metaphor used ("Perspective Clone Tool").

Rhetorical tropes.

Perhaps the most prominent of the rhetorical tropes in this context is the metaphor. As is the case of syntax, the same set of metaphors is shared with Adobe Photoshop.

Interaction phases.

The interaction phases are similar to those mentioned above in the Adobe Photoshop analysis. Also, the interaction sentence level is similar. However, the middle phase (where the user works on the picture) seems to be more consistent with beginning and end. This is because the user keeps working in the image window and is not distracted by other windows or palettes.

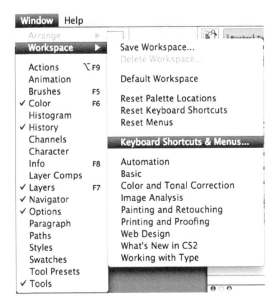

FIGURE 4.18 Adobe Photoshop customize UI. Source: Adobe Photoshop CS2. Adobe product screenshot(s) reprinted with permission from Adobe Systems Incorporated.

4.1.5 CUSTOMIZE THE UI

Adobe Photoshop steps

(1) Find the proper function in the menu or tool palette.

   (a) The subtask involved was to look through the menu items (especially in what seemed as most related: Photoshop -> Preferences, and Edit) for a relative command (it was found under Edit -> Keyboard Shortcuts. . . , and Window -> Workspace -> Keyboard Shortcuts & Menus. . .).

   (b) Alternatively look in the toolbox for a button resembling the intended action. Nothing was found. (A two-tabbed window showing Keyboard Shortcuts appears.) (See Figures 4.18 and 4.19.)

(2) Click on the triangle next to the menu item in the list to toggle the subgrouped commands.
(3) Click on the keyboard shortcut field to select it.
(4) Press the new keyboard shortcut.
(5) Press "OK" to save.

HE analysis

Visible interfaces/WYSIWYG.

The UI customization function wasn't present on the toolbar, and was only accessible through the menu bar. Since it is one of the advertised features, it should be accessible as readily as possible.

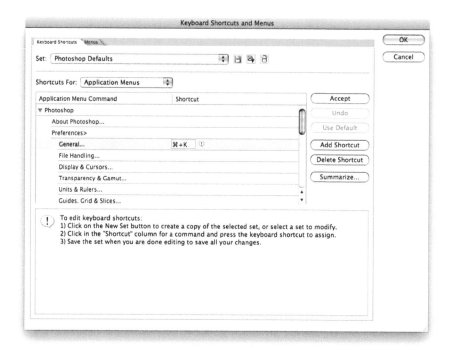

FIGURE 4.19 Adobe Photoshop customize keyboard. Source: Adobe Photoshop CS2. Adobe product screenshot(s) reprinted with permission from Adobe Systems Incorporated.

Modelessness; direct manipulation/see and point.

By selecting the UI customization function the user is presented with a new window (named "Keyboard Shortcuts and Menus") containing a list of menu items and keyboard shortcuts. After the adjustments the user has to click "OK" to transfer the changes to the UI. A better solution would be using a drag-and-drop interface, like the "Customize Toolbar. . . " in the Finder or other native Mac OS X applications. By doing so, we would also provide the user with more direct control.

SA analysis

Actors, audience, paradigm.

All of the objects involved in the interaction pertain to the leading paradigm of "window, icon, menu, pointing device" (WIMP). The paradigm is constituted by the menu bar, toolbars, main window containing the image, dialog windows, icons, and pointer. The paradigm is bound to the GUI metaphor. Adobe Photoshop is meant for professionals. This distinction of audience is manifested implicitly by the channel of distribution (commercial software) and explicitly in the marketing documentation (Adobe Photoshop's slogan reads: "The professional standard in desktop digital imaging" (Adobe, 2005)). The menu paradigm is constructed by combining noun-verb or verb-noun items, which seems deliberate (only one model should be chosen.) A more specific audience for this function consists of power-users and programmers.

Symbols.

The users are addressed by symbols pertaining to the user domain. In this case, the application icon and splash screen of Adobe Photoshop feature a colorful feather. The connotations are elegance, simplicity, and naturalness, which one would expect from a professional tool. What might break the expectation, however, is the historical usage of the image that symbolizes a writing pen. The other screens (and toolbars) are very compact and gray. The menus are only text-based, whereas the toolbar has only icons (with a textual label). The icons in the toolbar are related to their object in different ways but are connected to the prevailing metaphor and follow the application genre conventions.

Syntax.

The system processes are constituted by UI language components, as described earlier. In the interaction transcript, we can find all the elements mentioned. There are basic lexemes ("click," "option-click"), interaction sentences ("Select the function to assign to a different keyboard shortcut"), rhetorical tropes (e.g., metaphors, such as cards layout), interaction games (these are the complete functions enabling us to accomplish our goal, e.g., "Change a keyboard shortcut"). The designer's narration element is found in the tool-tip help reinforcing the icon meaning; in the status bar of the window or a dialogue window, it gives instruction on how to use the tool. Finally, in the Help menu it describes comprehensively the program functions. In the "Keyboard Shortcuts and Menus" window, the designer's narration gives detailed instructions for all the steps involved.

Rhetorical tropes.

Perhaps the most prominent of the rhetorical tropes in this context is the metaphor. The function metaphor is based on cards with labels, as in a filing cabinet. There are two cards: "Keyboard Shortcuts" and "Menus." The information in the window is mainly text-based. The window is split into regions of different type and functionality, which breaks the metaphor of cards.

Interaction phases.

On the level of interaction sentence, the interaction changes to reflect the constant evaluation of results on the user's part. The interaction sentence is then modified or repeated accordingly. Considering the example from the transcripts, the action is modified after the system's feedback (if we select a keyboard shortcut, that is already used somewhere else), the action is repeated (for different items in different menus), or the action was needed only once (when applying the changes by pressing the "OK" button).

GIMP steps

(1) Find the proper function in the menu or tool palette.

    (a) The subtask involved was to look through the menu items (especially in what seemed as most related: File and Edit) for a relative command (it was found under Edit -> Keyboard Shortcuts).

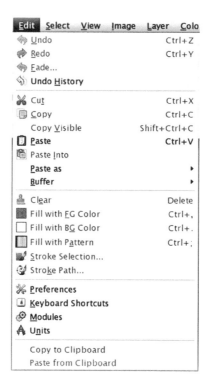

FIGURE 4.20 GIMP customize UI menu. Source: The GIMP 2.6.7.

    (b) Alternatively look in the toolbox for a button resembling the intended action. Nothing was found. (A "Configure Keyboard Shortcuts" window appears.) (See Figures 4.20 and 4.21.)

(2) Click on the triangle next to the menu item in the list to toggle the subgrouped commands.

(3) Click on the keyboard shortcut field to select it.

(4) Press the new keyboard shortcut.

(5) Press "Close" to save.

HE analysis

Visible interfaces/WYSIWYG.

The UI customization function wasn't present on the toolbar, and was only accessible through the menu bar. Since it is one of the advertised features, it should be accessible as readily as possible.

Modelessness; consistency and standards; direct manipulation/see and point.

By selecting the UI customization function the user is presented with a new window (named "Configure Keyboard Shortcuts") containing a list of menu items and keyboard shortcuts. After the adjustments the user has to click "Close" to transfer the

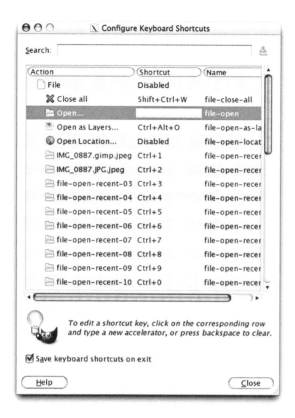

FIGURE 4.21 GIMP customize shortcut. Source: The GIMP 2.6.7.

changes to the UI. A better solution would be using a drag-and-drop interface, like the "Customize Toolbar..." in the Finder or other native Mac OS X applications. By doing so we would also provide the user with more direct control. Moreover, by using the text "Close" instead of "OK" or "Apply," the user may be unsure whether the changes will be applied.

SA analysis

Actors, audience, paradigm.

All of the objects involved in the interaction pertain to the leading paradigm of "window, icon, menu, pointing device" (WIMP). The paradigm is constituted by the menu bar, toolbars, main window containing the image, dialog windows, icons, and pointer. The paradigm is bound to the GUI metaphor. GIMP is intended for amateurs, semiprofessionals and programmers. This distinction is manifested implicitly by the channel of distribution (open-source) and explicitly in the marketing documentation (in GIMP it is by stating, that "[i]n the free software world, there is generally no distinction between users and developers" (GNU Project, 2009)). The menu paradigm is constructed by combining noun-verb or verb-noun items, which seems deliberate (only one model should be chosen). GIMP shares the same user group as Adobe Photoshop (power-users, programmers).

Symbols.

The users are addressed by symbols related to the user domain. In this case, Gimp's icon of a stylized dog head connotes playfulness, fun, and ease of use. The icon is not used on the splash screen, however, in favor of a planet picture. The toolbars and other screens show larger and more colorful button icons and larger dialog windows that are easily reached by the pointer. The icon symbols used in the menus, e.g., in the "Tools" menu, makes no distinction between nouns (e.g., Pencil, Eraser, Text) and verbs (e.g., Zoom, Measure, Heal), which could be helpful. Also, the symbols are created by different methods (e.g., the Pencil tool has an iconic representation of a pen but the Zoom tool icon is created by metonymy with its action and uses a zooming lens; other are connected only loosely, as in the case of Swap Colors).

Syntax.

The system processes are constituted by the same UI language components analyzed above for Adobe Photoshop.

Rhetorical tropes.

Perhaps the most prominent of the rhetorical tropes in this context is the metaphor. As is the case of syntax, the same set of metaphors is shared with Adobe Photoshop. The function metaphor is based on a list of actions. The information in the window is mainly text-based. The window is split into regions of different type and functionality, but the metaphor isn't contradicted anywhere.

Interaction phases.

The interaction phases are similar to those mentioned above in the Adobe Photoshop analysis. Also, the interaction sentence level is similar. Considering the example from the transcripts, the action is modified after the system's feedback (if we select a keyboard shortcut, that is already used somewhere else), the action is repeated (for different items in different menus), or the action was needed only once (when applying the changes by pressing the "OK" button).

4.2   GLOBAL SEMIOTIC ANALYSIS

Although we can analyze the different functions separately within a limited context by looking into the usage of icons, rhetorical tropes, and user expectations, SA can perform best when there is a larger UI corpus and as wide a context as possible. To allow for this context, we shall analyze first the domain of both the applications and then compare them in terms of paradigm, syntagm (syntax), rhetorical tropes, and codes.

Actors, audience, paradigm.

All of the objects involved in the interaction pertain to the leading paradigm of "window, icon, menu, pointing device" (WIMP). The paradigm is constituted by the menu bar, toolbars, main window containing the image, dialog windows, icons, and pointer. The paradigm is bound to the graphical UI metaphor. Both the applications analyzed belong to the same genre, as we have shown above.

Patterns.

The patterns formed from the interaction games are the following (based on the functions analyzed in the UI corpus): Open a document, choose a tool, apply a tool, save the file. The patterns from interaction sentences are centered on the verbs: find, look, select, click, drag, open, and save. The nouns are: tool palette, button, brush, object, menu, window. The patterns from the discrete elements are: mouse-drag, mouse-up, mouse-down, option-press, control-press.

In order to analyze the number of steps and patterns more easily, the UI language elements should be visualized in a future work.

Each step groups an interaction sentence with the system's response.

1. Red-eye removal: PS 5 steps, GIMP 4 steps.
2. Barrel distortion: PS 8 steps, GIMP 8 steps.
3. Eliminate an object: PS 7 steps, GIMP 10 steps.
4. Clone objects in perspective: PS 7 steps, GIMP 6 steps.
5. Customize the UI: PS 5 steps, GIMP 5 steps.

As is evident in the "eliminate an object" analysis, the number of steps can vary depending on the user's (mis)interpretation of the interaction proposed by the symbols used or based on the user's knowledge of the target domain or proficiency with the software. The number of steps can be thus reduced by using keyboard shortcuts or prolonged by consulting the manual during interaction.

Even on such a small set of five functions, different genres of interaction begin to emerge. So far we can distinguish two of them: in the first four functions the genre is image manipulation; in the last function it is UI modification. Obviously, the greater the number of functions analyzed, the stronger the evidence for our claim.

4.3   EXPERT EVALUATION RESULTS

By comparing the output from the HE and SA analysis, HE proved to be more concise. However, of the 16 heuristics used, only a small number could be applied on each occasion. The application of the 6 elements of SA tended to be more verbose, but, on the other hand, the elements could be applied every time. Whereas SA could seem repetitive in some instances, it provided a solid context of analysis. Both the methods (HE and SA) could be used not only on the interaction sentence level but also as for a general appreciation of the entire UI. During the general analysis, only portions of the UI are selected and suggestions made to other similar parts of the UI. In summary, our study demonstrated the depth of investigation and breadth of insight that SA can achieve in HCI and how this could enhance the current UX practice. Both methods could be merged to provide a best-of-both solution.

The changes that should be tracked and analyzed range from simple design iteration of the UI to more complex revisions, as is the case of the transposition from graphical UI to tangible UI, from desktop to mobile (or vice versa), or from one cultural background to another (e.g., when localizing a product).

Both the methods (HE and SA) could be used not only on the interaction sentence level but also as for a general appreciation of the entire UI. During the general analysis only portions of the UI are selected and suggestions made to other similar parts of the UI. Traditionally, these methods have been employed for goal-oriented work applications. However, they can be also applied to a range of different applications, including entertainment applications, where the goal is not always clearly defined, in a static or mobile setting.

5 Discussion

In Part I, we presented a semiotic approach to HCI/UX. From defining the basic semiotic theories that might used for such a purpose we then shifted to discuss the different elements of interaction language. We also presented semiotics as an analytic method especially in its most complex dimension—pragmatics. Pragmatics stands in the design process at the beginning because it forms the strategy and purpose of the developed UI. In the sign context, pragmatics leads the meaning interpretation—what semantics will be assigned to which syntax elements. Not only is this a process of interpretation, but also the whole UI development strategy is subject to a HCI ideology to a large extent. Such HCI ideology acquires its specific form in the UI. For the purposes of developing new UIs, and also for interacting with the UIs already in place, it is important to know the ways in which pragmatics, as an interpreting principle, is coded and mediated. We can then counter the ideologies by proper education and analysis.

A solution of how to leverage such a situation is therefore, on one hand, maximizing one's competence in terms of coding forms and medialization that has a big impact on the creation of UI. On the other hand, there is a need to develop methods analyzing the influence of such UI on the society, the creation and modification of meaning, and human relations that would be able to uncover the design behind every design. This is in line also with Fogg's suggestion:

> One useful approach is to conduct a stakeholder analysis, to identify all those affected by a persuasive technology, and what each stakeholder in the technology stands to gain or lose. By conducting such an analysis, it is possible to identify ethical concerns in a systematic way.

(Fogg, 2003, p. 233)

In the UI corpus (Section 3.3) we presented the transcript of interaction sentences forming language games, that served together with the actual UI as a basis for HE evaluation and SA analysis. Moreover, the transcript served as an input for defining the different elements involved in the interaction language.

And, in the case study we applied the semiotic and linguistic theories to an UI corpus of graphic manipulation applications. At the same time we analyzed the corpus using heuristic evaluation.

The SA provided the expected kind of data (e.g., conventions, connotations, combinations), that gathered a wider context than those from HE. That said, SA can be used to complement the widely used expert evaluation methods, but could possibly be defined to have a higher overlap with HE. In the latter case, SA would need to be evaluated hand in hand with the interaction sentences.

In summary, our study demonstrated the depth of investigation and breadth of insight that SA can achieve in HCI and how this could enhance the current UX practice. Both methods could be merged to provide a best-of-both solution.

Part II

Culture of Interaction

Parts of this section are used with the kind permission from Springer Science+Business Media: Cross-cultural comparison of UI components preference between Chinese and Czech users In: HCI International 2013 Conference Proceedings by Springer in the Lecture Notes in Computer Science (LNCS) series. The original publication is available at http://www.springerlink.com.

6 UI Language Components and Cultural Markers

Given the intensification of globalization through communication technology, we are faced more and more with UIs coming from different cultural backgrounds. There is also a growing need to design UIs that are usable and well accepted in a targeted culture. In order to match the user's cultural expectations as closely as possible, designers need to combine usability knowledge with cultural insights to form a "culturability," as coined by Barber and Badre (1998). Cross-cultural testing of UIs is the most comprehensive way to meet this goal, but it is also the most financially demanding. Therefore, by defining a usable set of UI design guidelines for a target culture, designers could market their products with lower costs and with better acceptance.

In our work we followed and expanded upon a body of previous research in the field of cross-cultural research (e.g., Choong and Salvendy, 1998; Clemmensen and Roese, 2010; Evers, 1998; Ge et al., 2007; Hofstede, 2010; Hotchkiss, 2007; Kurniawan et al., 2001; Marcus, 2001; Nisbett and Masuda, 2003; Shen, 2006; Smith, 2011; Tavassoli, 2002). Perhaps the most used model for cultural analysis is Hofstede's (2010), which works with the following cultural dimensions: power distance, uncertainty avoidance, masculinity vs. femininity, individualism vs. collectivism, and time orientation. In the field of cross-cultural comparison (Dong and Lee, 2008), we can build upon a body of previous research (Marcus and Gould, 2000; Sheridan, 2001; Smith et al., 2004). In our view, however, only limited work has been done in creating usable guidelines for cross-cultural UI design. We bring our insights from our cross-cultural work and propose a set of design guidelines. Our approach, however, seeks to find different categories that are directly based on our UI language components.

To promote this line of research, we decided to create a set of guidelines that could be used to enhance the user's acceptation of the UI in a specific culture. To acquire the necessary insights, we conducted a pilot study targeted at the habits, mental models, and UI preferences of Chinese and Czech users. For this purpose, we chose to work from the semiotic perspective that helps us uncover the sense-making processes of the users. We used semiotic methods to build a common framework to gather and analyze cross-cultural data. From our perspective, the UI is an example of complex language. Consequently, in our research we focused on different components of the UI language such as: discrete elements, interaction sentences, narration, rhetorical tropes, and patterns. Focusing on these UI language components allowed us to focus the scope of our research.

In order to focus the research, different types of signs were identified. These signs are taken from all the semiotic planes, namely pragmatics (e.g., trust), semantics (e.g., navigation controls), syntax (e.g., color combinations), and lexical (e.g., direction of written language). In the context of cross-cultural research these signs are called cultural markers (Barber and Badre, 1998) or cultural attractors (Smith, 2004). "Cultural markers are interface design elements and features that are prevalent, and possibly preferred, within a particular cultural group" (Barber and Badre, 1998, p. 2).

Cultural markers can be: "Colors, spatial organization, fonts, shapes, icons and metaphors, geography, language, flags, sounds, and motion" (Badre, 2001, p. 2). In a similar vein, cultural attractors "define the interface design elements of the website that reflect the signs and their meanings to match the expectations of the local culture. The cultural attractors typically comprise of: colours, colour combinations, banner adverts, trust signs, use of metaphor, language cues, navigation controls and similar visual elements that together create a 'look and feel' to match the cultural expectations of the users for that particular domain" (Smith, 2004, p. 1).

In our study we enhanced the UI language components with the cultural markers to obtain a more amenable framework for our cross-cultural research. The outcome is discussed in the following chapter.

7 Cross-Cultural Design and Evaluation Framework

7.1 RESEARCH METHODS

To compare the HCI/UX across different cultures we evaluated the following options:

- Run a usability testing between Adobe Photoshop and the GIMP. The results would enhance the expert evaluation done in Part I, "Semiotics of Interaction." On the other hand, some of the users may have been familiar with the applications, which would distort the results. The test setting would be quite complex, and the results may have not shown a very large scope of cultural markers, because both the applications come from the Western context.
- Run a usability testing of webpages from different cultural backgrounds. In this test we would gather more diverse results. However, some of the pages share the same design patterns across cultures, some of them are localized versions of Western web portals, which would limit our research scope.
- Run a usability testing of a UI prototype built (or modified) according to our proposed guidelines. This would be perhaps the most time-intensive and risky approach to the research. Risky, because we would build on top of guidelines that we did not evaluate before.
- Conduct an ethnographic observation and user interview. This kind of research would gather a large scope of cultural insights, but not all of them would be comparable between cultures. Also, we would not be able to focus on many cultural markers.
- Validate both previous research results and newly generated hypotheses through usability testing and interviews. In this way we would create a ground, from where we could build our guidelines. Validating the results and hypotheses does not allow delving very deep into different research subjects, but would allow us to cover a large ground, while working with a larger sample. We chose this kind of method for our pilot study.

In order to find the prevalent and preferred UI language components and cultural markers, we focused our study on the five following areas: personal information (demographics, exposure to other cultures and technologies), layout (discrete elements, patterns, interaction sentences and narration), color (discrete elements, rhetorical tropes), symbol (rhetorical tropes), and look and feel (interaction sentences, narration, patterns, and rhetorical tropes). There were few overlaps due to the broad scope we focused on in this pilot study.

Our qualitative method was based on one-to-one and one-to-many interviews supported by note taking and filling in questionnaires. In order to test the different hypotheses, we created at least one question for each hypothesis. The questions were clustered according to their areas, that is, personal information, layout, color, symbol, and look and feel. When appropriate, we used closed questions. However, because we carried an exploratory pilot study, we did not want to constrain our respondents and offered mostly open questions. The questionnaire was supported by examples of UI components.

To get data as reliable as possible, we wanted to limit the respondent's adaptation to a foreign culture. According to Smith (2011), tests should be done locally to prevent users from absorbing the majority culture, and should be carried on with a local moderator, sharing the language and cultural background (Ibid., p. 33). For that reason, we worked with students who were enrolled at a local university (Dalian Maritime University in China and Charles University in Prague, the Czech Republic) and were born and lived in the target cultures of our study. Also, the moderator of the interviews was a native speaker at each location.

For each of the interviewed groups we chose a sample consisting of 20 respondents, evenly split between females and males. The Chinese respondents had a mean age of 23 years, while the Czechs had a mean age of 26 years. For this kind and depth of research, 20 respondents make for an adequate saturated sample size (Guest et al., 2006; Nielsen and Landauer, 1993; Mason, 2010). The results were analyzed both qualitatively and quantitatively with regard to the threshold of significance. The threshold of significance was set to 10%. Only results of more than 10% from one another were taken as different. This level is based on Sauro (2005), where there is an 8% margin of error for results from 20 respondents. We took into account only results greater than or equal to 60%. The open questions were then analyzed with content analysis and contextual analysis across multiple questions where appropriate.

The research was carried on in the following manner:

- Users were presented with a localized **questionnaire** of 53 questions divided into groups: personal info, layout, color, symbols, look and feel.
- The questionnaire was supported by a **moderator**, who would give the necessary background to the user, and present him or her a choice of images/color samples, when appropriate.
- The testing of the Chinese sample was carried out in the **user lab**, with a **note taker** sitting besides the moderator. The Czech sample was interviewed in a **computer lab**, and the respondents answered the questionnaire directly by typing.
- In the user lab, users were **taped** by 3 cameras and their voices were recorded.
- Supportive **note taking** (2–3 people), and/or observation were done in an adjacent room.
- As a data-collecting and reporting platform, we chose `SurveyGizmo.com`.
- One session took about **90 min**.
- After each testing day, data from the note takers was **checked and merged** with the data on the server.

- Later on, the data was translated into English for analysis and presentation.

7.1.1 PHASES OF RESEARCH

- Defined research **questions**
- Defined **hypotheses** about the Chinese users
- Built a **questionnaire** of ~50 questions for user interviews using an online reporting tool (`surveygizmo.com`)
- Wrote a recruiting **screener** and set up an online recruiting form
- Wrote a recruiting **letter** and distributed it through social networks and e-mail conferences
- **Interviewed** 40 respondents

The respondents of our pilot study were screened according the following parameters:

- Born and raised in China/Czech Republic
- Belong to the Chinese/Czech culture (have Chinese/ Czech parents/mother tongue)
- Evenly split between males and females
- Evenly split between bachelor's and master's students
- Evenly split between living in large cities and the countryside
- Have different computer knowledge levels
- Have different English language levels

The results of the screener are contained in the "personal" section of the questionnaire.

7.1.2 INITIAL QUESTIONS

We started with the following list of research questions:

Q1: How does the level of English language proficiency alter the perception of UI/task structure?
Q2: How does the level of knowledge of computer use alter expectations from the UI interaction?
Q3: What are the language (and/or cultural) particularities that help or hinder successful user interaction?

Given the limited scope of the pilot study, we focused on the latter question, and revised the questions as follows:

Q1: Can we construct a set of design guidelines for Chinese users from what we know about the East/West differences?
Q2: Are there any culture-specific HCI/UX patterns to consider?

7.2   HYPOTHESES AND QUESTIONNAIRE

For the pilot study, we gathered 45 hypotheses about the specifics of Chinese users. The hypotheses were drawn from the conclusions of previous research in usability testing, psychological studies, visual semiotics, and linguistics. For the information value of UI components's spatial organization, we worked with the oppositions of Given/New, Ideal/Real, and Center/Margin proposed by Kress and van Leeuwen (2006). In this context Given is taken-for-granted information and New information is introduced later. Similarly, Ideal presents what might be, and Real what is, for example, a specific or practical information. All of these oppositions can be combined with Center, a nucleus of information (or the most important information), and Margin, containing other dependent information.

Some of the conclusions were directly included in the hypotheses (e.g., favorite colors for the background: blue, purple, cyan, gray), some were modified according to our assumptions (e.g., because Given information in the West is expected on the left of a screen, we expected the information to work better on the right in China), some were constructed from our direct experience with the Chinese culture (e.g., red color with yellow text is used for special occasions), while some tested our more general assumptions (e.g., there is a close similarity between the sequential information structure in language and the horizontal structure in visual composition).

Some of the hypotheses that came from the above process of construction did not perform well in the early tests, or were too vague to translate into a question, that would be meaningful in the context of the study. Such hypotheses and the related questions were omitted from the research. The complete list of hypotheses is available in Appendix C, "Hypotheses."

In order to test the different hypotheses, we created at least one question for each hypothesis. The questions were grouped according their areas, that is, personal information, layout, color, symbol, and look and feel. Where appropriate, we used closed questions. However, because we carried an exploratory pilot study, we did not want to constrain our respondents and offered mostly open questions. The questionnaire was supported by examples of UI components.

8 Case Study: Cross-Cultural Preference of UI Language Components, Cultural Markers

8.1 INTERVIEW AND RESULTS ANALYSIS

Our findings show there is a strong influence of globalization on the cultural markers, mainly through the use of common software platforms. In spite of that, we found still many important culture-specific differences in both groups that are related to: spatial organization of information (Kress and Van Leeuwen, 2006), shapes, direction of reading, motion, color, color combinations, semantic organization of content, use of icons and metaphors, user's preferences for different types of media, preference for culture-specific content and for cartoon imagery, trustworthiness of the content, navigation tools, and visible and interaction grammar of menus and commands. Almost half (22) of our hypotheses were fully supported by the results from individual questions, 17 were partly supported (e.g., the result came second with a small difference in percentage after the first answer), 14 hypotheses were not supported, although useful information could be extracted, and 2 hypotheses were impossible to verify due to lack of data. In the following subsections, we provide a summary of the hypotheses that were supported by the data, those that were not, as well as other interesting insights and comments. The summary is divided by the main themes of our research.

8.1.1 PERSONAL DATA

Our population sample consisted of 20 respondents, evenly split between females and males. The Chinese respondents had a mean age of 23 years, while the Czechs had a mean age of 26 years.

1) How old are you?

Results summary.

The majority of the Chinese respondents (76%) were aged between 22 and 23 years. Their average age was 23 years. The majority of the Czech respondents (55%) were between 19 and 26 years of age. Their average age was 26 years. (See Figure 8.1.)

2) What is your gender?

Results summary.

Both Chinese and Czech respondents were split exactly in half between males and females. (See Figure 8.2.)

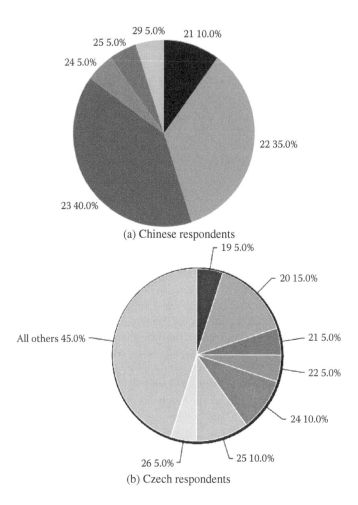

FIGURE 8.1 Age of respondents.

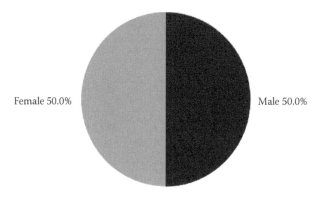

FIGURE 8.2 Gender of respondents.

3) Where were you born?

Results summary.

The vast majority of the respondents were born and raised in the same place.

Discussion/conclusion.

The rationale of the question (together with other ones) is to make sure the respondents come from the local culture and were exposed to it from an early age.

4) Where were you raised?

Results summary.

The vast majority of the respondents were born and raised in the same place.

Discussion/conclusion.

The rationale of the question (together with other ones) is to make sure the respondents come from the local culture and were exposed to it from the early age.

5) Where do you come from?

Results summary.

60% of Chinese respondents came from rural areas, whereas 40% from urban areas. 85% of Czech respondents came from urban areas, and 15% from rural areas. (See Figure 8.3.)

6) What is your mother tongue?

Results summary.

The mother tongue of all the Chinese respondents was Mandarin. The Czech sample contained two Slovak respondents.

Discussion/conclusion.

The rationale of the question (together with other ones) is to make sure the respondents come from the local culture and were exposed to it from an early age.

7) Which languages do you speak?

Results summary.

All of the Chinese respondents spoke both Mandarin and English. Two of them (10%) also spoke Japanese. In the Czech sample, all of the respondents spoke English, 10 of them (50%) also German, and six also Spanish (30%).

Discussion/conclusion.

The respondents were all exposed to the English language, although there were large differences between their knowledge.

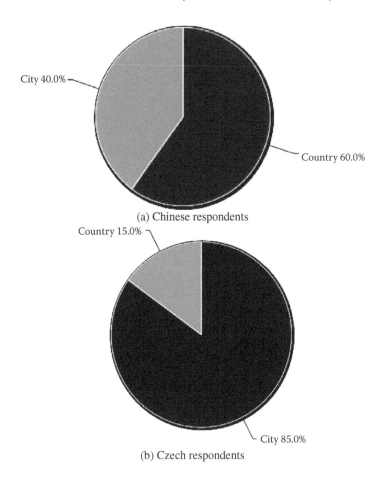

(a) Chinese respondents

(b) Czech respondents

FIGURE 8.3 Origin of respondents.

8) What is your English language level?

Results summary.

Almost half of the Chinese respondents (45%) stated they had an upper-intermediate level of English (corresponding to the B2 level according to the European framework), 35% declared an advanced level (C1), and 20% had an intermediate level (B1). In the Czech sample, 35% stated they had an intermediate level of English (B1), 35% stated they had an upper-intermediate level of English (or B2), and 30% declared an advanced level (C1). (See Figure 8.4.)

Discussion/conclusion.

Our intention was not to verify the level of English language per se, but to see whether there is a relation between the level of English knowledge and the understanding

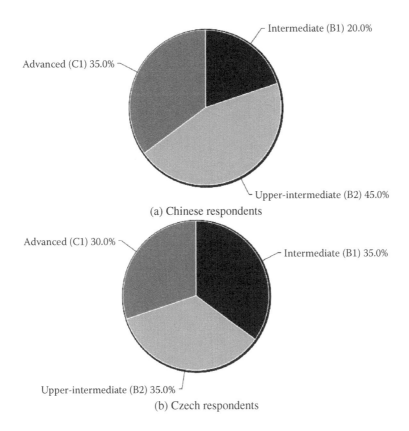

(a) Chinese respondents

(b) Czech respondents

FIGURE 8.4 Level of English language in respondents.

of foreign UIs. Our assumption is that the majority of UIs is inherently structured according to the English language.

9) What is your computer knowledge level?

Results summary.

30% of the Chinese respondents felt they had an intermediate level of knowledge, the same number stated they had an advanced level. Both elementary, and programmer level of knowledge gained 20% of all the respondents. In the Czech sample, 40% of the respondents felt they had an intermediate level of knowledge; the same number stated they had an advanced level. (See Figure 8.5.)

Discussion/conclusion.

The levels of knowledge were split quite evenly. Our assumption was that the higher the computer savviness, the lower the influence of the mother tongue structure on the user's interaction would be. We hold that the UIs are structured differently from the

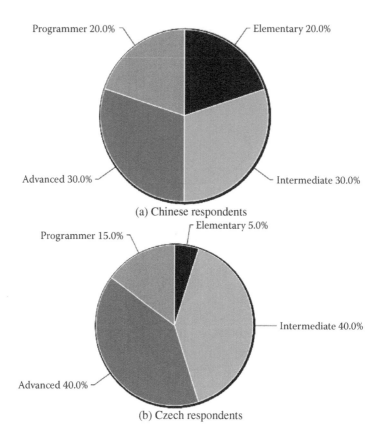

FIGURE 8.5 Computer level of respondents.

mother tongue, mostly according to the English language. Given the subjects of this study, however, the above assumption could not be proved.

10) What is your favorite software application?

Results summary.

The most favorite applications in the Chinese sample were QQ instant messenger (60%), Sina microblogging (20%), 360 antivirus (20%), and Microsoft Office (Word especially with a gain of 20%). Chrome browser, MATLAB research tool, Skype, and Storm player were also used. In the Czech sample, however, there was no significant preference for a software.

Discussion/conclusion.

The favorite applications were often cited by the respondents in other parts of the study.

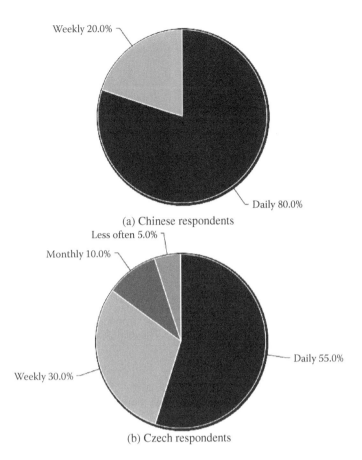

(a) Chinese respondents

(b) Czech respondents

FIGURE 8.6 Applications usage frequency.

11) Why?

Results summary.

The main motivations for using the applications were: simplicity, speed, ease of use, features, and low cost.

12) How often do you use it?

Results summary.

80% of the Chinese respondents used the applications daily, 20% weekly. However, only 55% of the Czech respondents used the applications daily, and 30% weekly. (See Figure 8.6.)

13) What is your favorite website?

Results summary.

The most popular websites for Chinese respondents were Renren social networking (40%) and Sina social networking and news (20%). Google search, Taobao shopping,

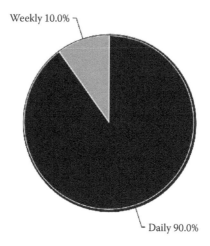

FIGURE 8.7 Website usage frequency.

123 news, and QQ messaging were also cited. The most popular websites in the Czech sample were Facebook (40%) and Gmail (25%).

Discussion/conclusion.

The favorite websites were often cited by the respondents in other parts of the study.

14) Why?

Results summary.

The main motivations for using the websites were: features, communication with friends, and for e-shops low postage.

15) How often do you use it?

Results summary.

The vast majority of both the Chinese and Czech respondents (90%) visited the sites daily, only 10% weekly. (See Figure 8.7.)

16) What is your e-mail address?

This question was asked for identification purposes and was excluded from this report for privacy reasons.

8.1.2 LAYOUT

For testing the UI composition we used a matrix with three rows and three columns (Figure 8.8). The hypotheses that were supported in relation to the spatial organization of the UI, shapes, direction of reading, and motion are:

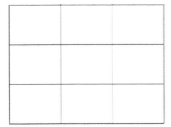

FIGURE 8.8 Layout matrix.

- Given information is expected on the right of the screen.
- A central composition is regarded more aesthetically pleasing than triptych composition.
- An even number of elements is more preferred than an odd number; ideal is eight.
- Images placed symmetrically in the middle look better than on the left or right of the screen.
- Free-flow layout is easier to use than grid-based layout.
- Users tend to read from top left towards the center of the screen.
- Left-to-right lines of text are easier to read than top-to-bottom and right-to-left.
- There is a close similarity between the sequential information structure in language and the horizontal structure in visual composition.
- Curves stand for softness (and would be better perceived), while straight lines for hardness.
- Rounded corners (curvilinear patterns) are better perceived than square corners (geometrical patterns).
- Copied UI elements are better perceived than original elements. This applies to both computer icons and design patterns.
- Icons presenting objects with a description are more understandable than those without a description.

The unsupported hypotheses, on the other hand, disclosed interesting details:

- Real information is expected on the bottom of the screen. The majority of Chinese respondents put real information in the middle level of the screen (middle row in the matrix), overlaying it thus partly on the new and ideal information.
- A square and double-square layout would be more preferred because it is widely used in Asia (a symbol of Earth, Japanese buildings, etc.). Instead, respondents preferred a golden-section layout, such as 16:9 or 4:3.

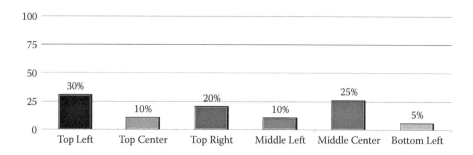

FIGURE 8.9 Placement of important information on the matrix in the Chinese sample.

17) On a screen, where would you place the most important information?

Chinese

Hypotheses.

- Users tend to attribute more importance on elements placed in the center of the screen.
- Users tend to read from top left towards the center of the screen.

Results summary.

The largest group of the Chinese respondents chose the top-left corner to hold the most important information (30%), because it is easily noticed and follows the prevalent reading pattern. Middle center was the second largest choice (25%), because it is usually the largest part of the screen, and holds most of the content. The third largest group chose the top-right placement (20%). The reasons came from habits (from some software or websites), as well as from reading patterns. (See Figure 8.9.)

Czech

Results summary.

The Czech respondents chose the middle center to hold most of the important information (40%) based on respondents' expectations. Also, the center is the first place where users look. Top left and top center followed second in popularity with 25% each. Top center was chosen because of its position in the height of the eyes, and also because it is located at the golden ratio. The other respondents chose top left because of reading habits. The choices were motivated by habit. (See Figure 8.10.)

Discussion/conclusion.

The hypotheses were partly supported by the results. The placement of important information follows the prevalent reading pattern. Therefore, the preference goes from the top-left corner towards the center. Also, the preference is related to the favorite websites and applications the respondents would use; that is, the top-right corner is used by some notes applications displaying memos in there. Interestingly, the Czech respondents preferred the uppermost row and the central column of the

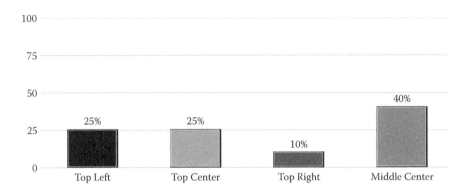

FIGURE 8.10 Placement of important information on the matrix in the Czech sample.

testing matrix. In contrast, the Chinese respondents favored mostly the uppermost row.

18) On a screen, where would you place new information?

Chinese

Hypotheses.

- New information is most readily noticeable in the top-left corner of the screen.
- New information (key, unknown) is expected on the left.
- Users tend to read from top left towards the center of the screen.

Results summary.

The largest group of the respondents would place new information in the middle center of the screen (35%). According to them, it is the place attracting their attention the most, and would look there first. Top-center placement was cited as second (25%), mostly because of users' reading habits, but also because they expect new information above the important information. The third largest group is top right (20%), and the main reasons are reading habits and habits from favorite applications. (See Figure 8.11.)

Czech

Results summary.

The largest group of respondents (40%) chose top right as the place for new information, mainly because of their habit from Facebook or Windows notifications. Also, such a placement does not interfere with the content of the window. Top-left placement gained 20%, because it is easily noticeable given the top-to-bottom reading direction. Top-center placement gained also 20%, the reasons being reading direction, but also habit from webmail and Facebook. (See Figure 8.12.)

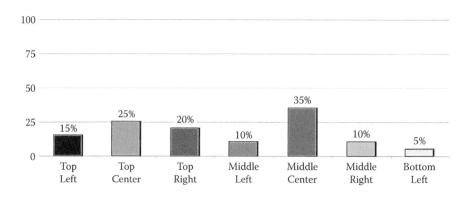

FIGURE 8.11 Placement of new information on the matrix in the Chinese sample.

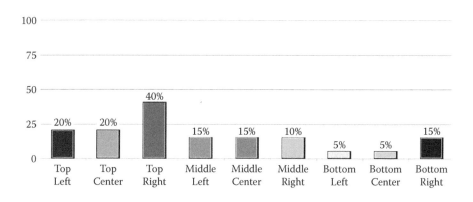

FIGURE 8.12 Placement of new information on the matrix in the Czech sample.

Discussion/conclusion.

The hypotheses were partly supported by the results. In the Chinese sample, the top-left corner came only second, contrary to our expectation. The top-right corner with new information follows the Western expectation, according to Kress and Van Leeuwen (2006). Interestingly, Chinese respondents preferred a placement of new information in the uppermost row and central column, whereas Czech respondents favored a placement in the rightmost column and in the uppermost row.

19) On a screen, where would you place familiar, taken-for-granted, unproblematic information?

Chinese

Hypothesis.

Given information (familiar, agreed upon) is expected on the right of the screen.

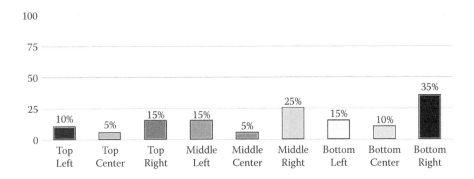

FIGURE 8.13 Placement of familiar information on the matrix in the Chinese sample.

Results summary.

Familiar information would be placed by the largest group of the respondents (35%) to the bottom-right corner of the screen. The main reason is that familiar information is not to be used so often, so it can yield to more important and more often used content. Also, the positioning was influenced by users' habit of keeping their minimized applications. For choosing middle-right positioning, the second largest group (25%) argued they do not need to watch the familiar information so frequently, and are accustomed to the position through software applications. The third largest groups (each with 15%) were divided into top right, middle left, and bottom left. The given reasons were based on low importance of the content and habit from websites and applications. (See Figure 8.13.)

Czech

Results summary.

The largest group of respondents (50%) would place familiar information in the top-right corner of the matrix, partly because of habit, but also, because it does not interfere with more important information. Middle-left placement came second (40%), because it feels natural/logical that way, and it does not interfere with the main action, which is supposed to take place in the middle of the screen. Also, habit is cited as a motivation. Third, came both top-left and bottom-left placement (30%), reasons being mainly that such information would shift there once read/accepted, as in the case of a mailbox. Also, the location right below the logo/header receives usually less attention than the logo or the center of the screen. (See Figure 8.14.)

Discussion/conclusion.

The hypothesis was supported by the results. The placement is opposite to that expected in the West, according to Kress and Van Leeuwen (2006). Interestingly, given information was placed by the respondents mostly in the lower half of the screen, where real (concrete, practical) information would be placed (Ibid.). This was underscored by comparing with the Czech results, where the preferred placement was in the uppermost row and leftmost column. The rightmost column was chosen both by

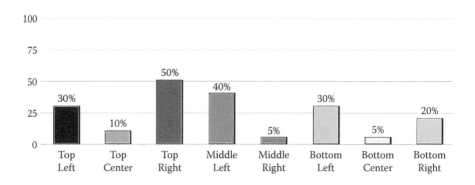

FIGURE 8.14 Placement of familiar information on the matrix in the Czech sample.

the Czech and Chinese respondents, while only the Chinese favored the lowermost row in the matrix.

20) On a screen, where would you place general information?

Chinese

Hypothesis.

Ideal information (symbolic, general) is expected on top of the screen.

Results summary.

General information would be placed most often in the middle-left (40%) part of the screen. As a reason, website navigation links are placed there, and while still in reach, they do not interfere with the content. Top left was the second most-cited placement (30%); it would be used for navigation, search box, as the top row is easily reachable, besides being a habit for most of the users to look for such information there. Bottom left, together with top right, was the third largest group of responses (15%). These placements were chosen mostly, because they do not interfere with other content, and may follow the habit from favorite websites. From other answers it turned out the right column is more usable for people accessing the content with their right hand. Interestingly, a logo would be expected in the top-right corner. (See Figure 8.15.)

Czech

Results summary.

The most popular placement was top right (35%), because of aesthetics and favorite applications and proximity to important information. Second came both top left and middle center (25%), reasons being that such information could expand in any direction and also because of expectations. Third (20%) were top center, middle right, bottom center, and bottom right. Besides the reasons already cited, bottom placement was chosen because it is on the margin and not needed for the main activity. (See Figure 8.16.)

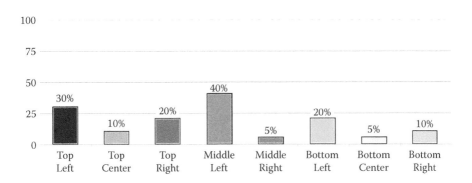

FIGURE 8.15 Placement of general information on the matrix in the Chinese sample.

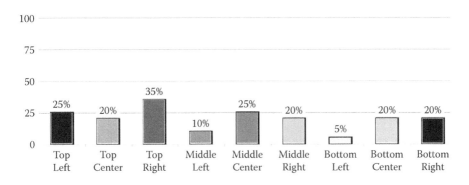

FIGURE 8.16 Placement of general information on the matrix in the Czech sample.

Discussion/conclusion.

The hypothesis was partly supported by the results. Most of the respondents put general information in the middle-left or top-left corner of the screen. Also, the left part of the screen was more preferred. Interestingly, this placement follows the given information, according to Kress and Van Leeuwen (2006). The Czech results showed a preference for the topmost row of the matrix, but also for the middle and rightmost column.

21) On a screen, where would you place detailed information?

Chinese

Hypothesis.

Real information (details, concrete image) is expected on the bottom of the screen.

Results summary.

From the majority of answers (70%) middle center would be the place to hold detailed information. The reason is its prominence in the attention field of the users, and habit also. Even more details would be expected to unfold towards the bottom. The

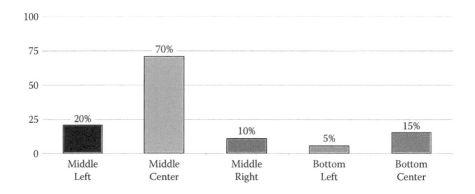

FIGURE 8.17 Placement of detailed information on the matrix in the Chinese sample.

second-largest group of answers (20%) shows a preference for the middle left, where according to the answers it gets much attention and is comfortable to read. The right side can be thus left blank. (See Figure 8.17.)

Czech

Results summary.

Most of the respondents chose bottom center (45%) to hold detailed information, in particular because it expands the main information located in the middle of the screen. Middle center followed with 40%, and middle right with 35%. The common reasons were logical relation and habits. (See Figure 8.18.)

Discussion/conclusion.

The hypothesis was not supported by the results. The majority of Chinese respondents put real information in the middle level of the screen (middle row in the matrix), overlaying it thus partly on the new and ideal information.

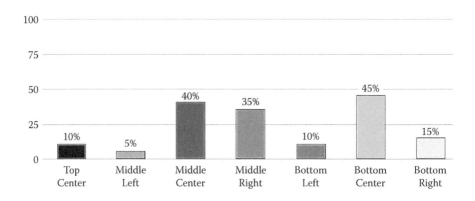

FIGURE 8.18 Placement of detailed information on the matrix in the Czech sample.

From the information placement questions it turns out that the new information would be placed in the upper half of the screen and in the middle, whereas given information would populate the middle level, but mostly the lower half. Such placement confirms the prevalent mutual exclusiveness of such information. However, in contrast to the left-right axis described by Kress and Van Leeuwen (2006), it follows the top-bottom axis. The ideal information would be mostly confined to the top-left quadrant, while the real would be placed in the middle left and center. The polarity of the latter two kinds of information is not so strong. However, while the real information is connected to the central part of the screen, the ideal information tends to populate the margins. Such a placement follows a central composition contrasting (as in a mandala with the Earth in the middle and Heaven on the margins) with the top-bottom axis suggested (Ibid.) for the Western (triptych) composition.

The Czech results showed also a preference for the center, but the strongest was towards the central row of the matrix.

22) On a screen, where would you likely put images?

Chinese

Hypothesis.

Images placed symmetrically in the middle look better than on the left/right of the screen.

Results summary.

Images were most often put in the middle-left part (30%) of the screen, because it is aesthetically pleasing and straightforward. Top right was the second largest group of answers (25%). The respondents were mostly concerned about the best combination with text. Also, pictures in the top row, if interesting enough, would lead the user to the text beneath. Middle center came third (20%), where images would get most of the attention. Also, they could be surrounded by text, but the text would look better to the left from the image. (See Figure 8.19.)

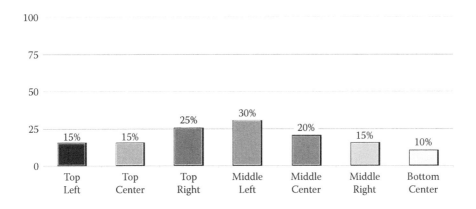

FIGURE 8.19 Placement of images on the matrix in the Chinese sample.

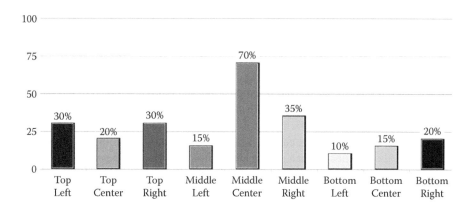

FIGURE 8.20 Placement of images on the matrix in the Czech sample.

Czech

Results summary.

Middle center came decisively first (70%), mostly because images usually form the main message of the page and should be most readily visible. Second came middle right (35%), because they should have space around them and be seen without scrolling. Third came both top-left and top-right placement (30%); if the image is not the main message, it is elegant and does not interfere with the main information. (See Figure 8.20.)

Discussion/conclusion.

The hypothesis was partially supported by the responses. The middle-center placement in the Chinese sample came only third, and a preference was given to an off-center positioning. Interestingly, the placement corresponds with real information placement from the above questions, as well as with new and ideal information (top-right corner). The strongest preference was for the middle row in the matrix. While the Czech results showed also a strong preference for the middle row, they demonstrated support for the uppermost row, the middle and rightmost column of the matrix.

23) Where is the first thing you usually notice on a screen?

Chinese

Hypotheses.

- There is a close similarity between sequential information structure in language and horizontal structure in visual composition.
- Users tend to read from top left towards the center of the screen.
- New information is more readily noticeable in the top-left corner of the screen.

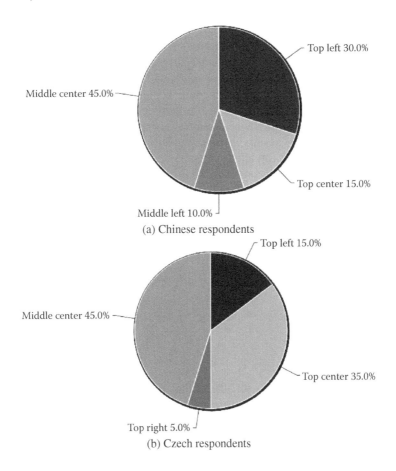

FIGURE 8.21   Placement of the first attractor on the matrix.

Results summary.

Most often, respondents would pay attention to the middle center of the screen (45%). Because of this habit the user would look there for the most important or new information. Top left came second (30%), the reason being that this is the starting point when scanning and reading the screen. (See Figure 8.21(a).)

Czech

Results summary.

The first place the Czech respondents noticed was middle center (45%), followed by top center (35%), and top left (15%). Middle center was a clear favorite, because it usually holds the main information. Also, the eye is attracted to the golden section of the screen, which is located between the first and second row, and the first and second column of the matrix. Top center was preferred mostly because of a habitual placement of navigational elements. Top left was chosen because of the reading direction, and incoming e-mail and logos are usually located here. (See Figure 8.21(b).)

Discussion/conclusion.

The hypotheses were partly supported by the results. Although previous research (e.g., eye-tracking, heatmaps) shows that the first part the user looks at is located in the top-left corner of the screen, our data support it from 30% only. This difference might be due to the method employed, because we were not measuring the subliminal reaction of the users, but were asking about their afterthoughts. Conversely, the strongest results spoke for the middle column, and this was shared between both the Czech and Chinese respondents. The results were shared between the Chinese and Czech results concerning the first place (middle center); however, the second and third locations were inverted (top left was second in the Chinese sample, but only third in the Czech sample. Top center came second in the Czech sample, and third in the Chinese one).

24) What is the first thing you usually notice on a screen?

Chinese

Hypotheses.

- The context of information presented is more salient than the information itself.
- Users would notice first the image, then titles, then the body of text.

Results summary.

The first things to be noticed with 50% of the responses were images, because they are intuitive and subconscious. Websites with images were regarded more trustworthy than text-only pages. The second largest group of answers (35%) indicated titles as the first element to be noticed on the screen. The main reason is that it summarizes the page content and provides thus a feedback for the user, if it is the page he or she was looking for. Interestingly, third came background (10%) that provides contextual information on the kind of site the user accessed (e.g., with red background the user would expect a site about the Communist party). (See Figure 8.22(a).)

Czech

Results summary.

In the Czech sample, both layout and image acquired 30%. Layout was chosen because the user needs to figure out where is what type of content and what does the screen mean. Image was cited, because it can convey the screen's meaning quickly, but also because images are usually the most attractive. Title came second with 15%, because it is usually the first information to load on a webpage, and it gives immediate feedback on the kind of site to the user. (See Figure 8.22(b).)

Discussion/conclusion.

The hypotheses were partly supported by the results. Only 15% of the Chinese respondents mentioned contextual cues (such as background or layout). The rest focused on the information presented. Future research could help by focusing on the duality between information given and information left out. The sequence of attention was

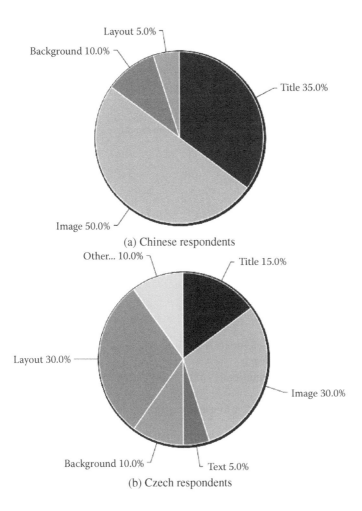

FIGURE 8.22 Kind of the first attractor on the matrix.

confirmed by the data. On the other hand, image gained the most votes both in the Chinese and Czech samples and title came second. However, Czech respondents favored layout (it shared the first position together with images), whereas for the Chinese respondents layout was the least important.

25) Which layout do you like the most?

For this question, we tested the layouts presented in Figure 8.23.

Chinese

Hypotheses.

- A central composition is regarded more aesthetically pleasing than triptych composition.

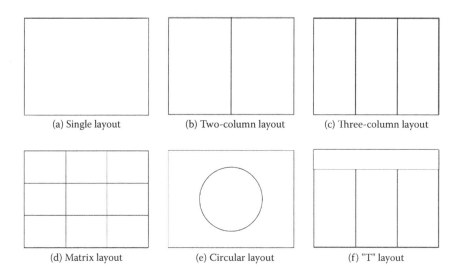

(a) Single layout   (b) Two-column layout   (c) Three-column layout

(d) Matrix layout   (e) Circular layout   (f) "T" layout

FIGURE 8.23 Tested layouts.

- An even number of elements is preferred more than an odd number; ideal is eight.
- Images placed symmetrically in the middle look better than on the left/right of the screen.
- Free-flow layout is easier to use than grid-based layout.

Results summary.

First, we tested a T-layout option, which started to be widely popular, in our view, due to its adoption in many favorite websites by the respondents. Therefore, we split the question into a T-layout and a three-column layout. The largest group of responses were given to the three-column layout (35%) and the T-layout (15%). Here, the navigation would be put on top, the most important in the left column, the less important in middle, and the least important in the right column. Alternatively, navigation and personal information on the left, images on the right, new information above the center, important information below the center. Third place (15% each) came to a two-column layout and central layout. With the former, respondents would put important information to the left and less important information or images to the right. The reasons for choosing the latter layout would be aesthetics and originality. The center would be used for important information or images. (See Figure 8.24(a).)

Czech

Results summary.

The most preferred layout had three columns (45%), followed by both the central and matrix layouts (20%); two-column layout came third (15%). The three-column layout was chosen, because it allows placement of the most important information in

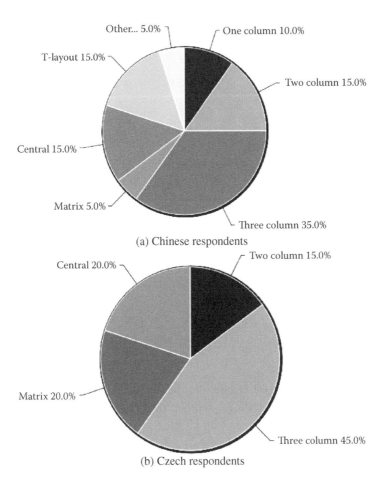

(a) Chinese respondents

(b) Czech respondents

FIGURE 8.24 Layout preferences.

the center, auxiliary information in the margin. It is traditional and usable. Central layout was chosen for its possibility to structure information as center and margin. Matrix layout was the favorite, because of its ability to structure information without clutter. Two layouts can concentrate the user's attention and structure information for navigation and content. (See Figure 8.24(b).)

Discussion/conclusion.

On the whole, the hypotheses were supported by the results. All of the layouts allowing for a central composition (three-column, T-layout, one-column, central) gained major support among the respondents. However, we could not assess the preference for the central composition (present in Asian art and culture, i.e., a mandala) vs. the triptych composition (known in the West from art and religion). Also, even though the grid-based layout gained small support, we could not verify the hypothesis that a grid-free (or free-flow) layout would be easier to use than the former one. The sequence of

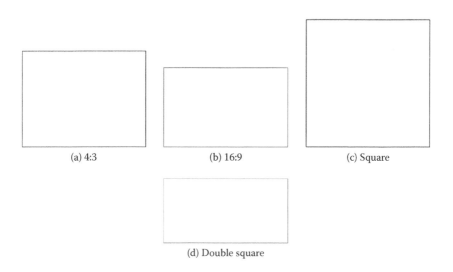

(a) 4:3   (b) 16:9   (c) Square

(d) Double square

FIGURE 8.25 Layout proportions preferences.

preference was shared between the Czech and Chinese respondents. The layouts allowing for a central composition gained a similar number of votes from both of the tested groups.

26) Which window proportions do you like the most?

For this question, we tested the layout proportions presented in Figure 8.25.

Chinese

Hypothesis.

Square and double-square layout would be more preferred because they are widely used in Asia (the symbols of Earth, Japanese buildings).

Results summary.

Exactly half of the users chose the 4:3 layout. Their motivations were based on habits (the respondents were used to the usual screen proportions), aesthetics, but also on practical reasons (they felt this proportion would display best most of the content). The 16:9 layout was chosen by 35% of the respondents, mainly because it is suitable for movies. Interestingly, about the square layout, with 15% of the answers, the respondents felt the most strongly. It was chosen for its contained size and ease. On the other hand, others regarded the square as having no focus, being too formal, too basic, or even depressive. The double square was regarded as generally unfit. (See Figure 8.26(a).)

Czech

Results summary.

The 16:9 proportion came first (50%), followed by 4:3 (40%). The first was chosen, because it can present information aesthetically, even though the content is not always

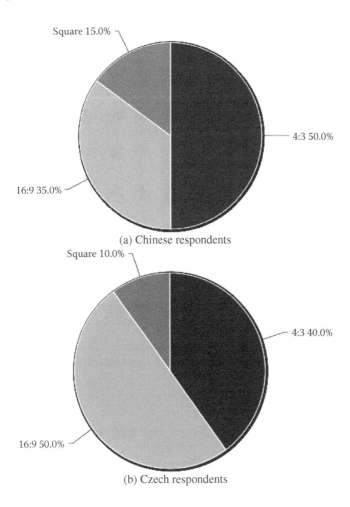

Square 15.0%

4:3 50.0%

16:9 35.0%

(a) Chinese respondents

Square 10.0%

4:3 40.0%

16:9 50.0%

(b) Czech respondents

FIGURE 8.26 Preference of layout proportions.

laid out optimally. The 4:3 layout was perceived as closer to the golden ratio. On the other hand, the square was not perceived well. (See Figure 8.26(b).)

Discussion/conclusion.

The hypothesis was not supported by the results. The majority of the users preferred proportions based on the golden ratio (such as 4:3 or 16:9), rather than strict geometrical ratios. There was only a small difference between Czech and Chinese respondents.

27) Which direction would you consider reading the text?

Chinese

Hypotheses.

- Users tend to read from top left towards the center of the screen.

FIGURE 8.27 Ideogram matrix to discover reading patterns. The characters are chosen with a similar number of strokes, so that all of them look very similar at first sight and do not attract the reader's eye.

- Left-to-right lines of text are easier to read than top-to-bottom and right-to-left.
- There is a close similarity between sequential information structure in language and horizontal structure in visual composition.

Results summary.

70% of the respondents preferred reading by lines from the left, as this is the prevalent reading habit learned from school and used also when scanning or searching some specific information. However, 30% of the respondents used other strategies when scanning a screen, such as starting from the center or from the bottom-right corner, then continuing in horizontal or vertical lines. (See Figures 8.27 and 8.28.)

Czech

Results summary.

Searching by lines was the choice of 25%; however, other strategies as a whole gained 60%. (See Figure 8.29.)

Discussion/conclusion.

The hypotheses were supported by the results. Interestingly, reading by lines (left to right) was the single largest strategy in the Czech sample, but only formed a quarter

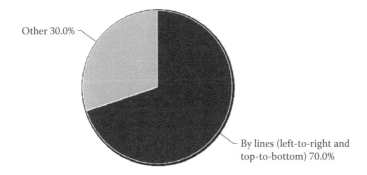

FIGURE 8.28 Reading direction in the Chinese sample.

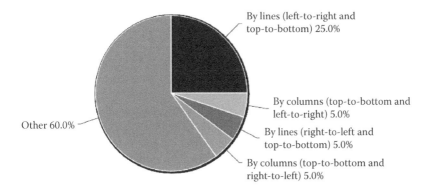

FIGURE 8.29 Reading direction in the Czech sample.

of the total. The Chinese respondents seem thus more disciplined when searching patterns, as they follow learned rules. It is important, however, that for the Chinese respondents the image represented a set of known characters (although in nonsensical sequences), whereas for the Czech sample they were mere shapes.

28) On a screen, if you saw the first image of a series only, where would you expect to find the next one?

For this question, we used the image in Figure 8.30.

Chinese

Hypotheses.

- Users tend to read from top left towards the center of the screen.
- Left-to-right lines of text are easier to read than top-to-bottom and right-to-left.
- There is a close similarity between sequential information structure in language and horizontal structure in visual composition.

FIGURE 8.30 First image of a series.

Results summary.

Exactly half of the respondents expected the next image to appear rightwards. The response was motivated by habits (direction of reading, going through a book, or websites). 25% of the respondents would expect the image from the left, also citing habits (reading direction and natural feel). 20% then would expect the image to come from the bottom, which would fit the reading direction starting from the top of a page. (See Figure 8.31.)

Czech

Results summary.

The most expected direction of showing next images in a row was right (80%), followed by bottom (20%). Right was a favorite answer because it follows the learned

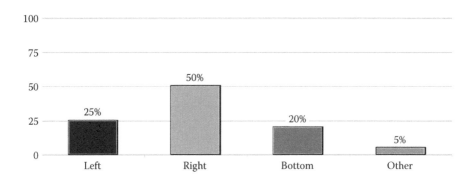

FIGURE 8.31 Next image expectation in the Chinese sample.

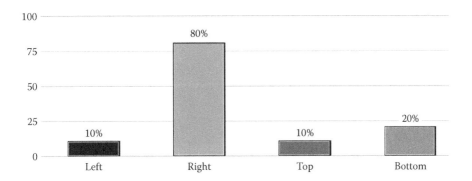

FIGURE 8.32 Next image expectation in the Czech sample.

reading direction; the respondents also expected the narration to develop towards the center of the screen. Also, a horizontal direction of development was more preferred, because it retained the eye level at the same height. The bottom direction was chosen because it followed the scrolling pattern on a webpage. (See Figure 8.32.)

Discussion/conclusion.

The hypotheses were partially supported by the results. The prevalent reading direction in the Chinese sample seems to have a strong influence on the expectation of dynamics of a visual composition, that is, a visual narrative. Although the vast majority of respondents grounded their choice on a natural reading direction, there were pronounced differences between them. These results should be triangulated with the data from similar questions. Bottom and left directions rated second in both of the groups.

29) Which of the following pair of items would you prefer?

For this question, we used the three sets of pairings presented in Figure 8.33.

Chinese

Hypotheses.

- Curves stand for softness (and would be better perceived), while straight lines stand for hardness.
- Rounded corners (curvilinear patterns) are better perceived than square corners (geometrical patterns).

Results summary.

The most favorite shapes with 45% of the answers had radius features. More than half (52%) preferred icons, windows coming second (35%), and dividers third (22%). Radius shapes were regarded as most balanced, soft, comfortable, and beautiful; also, most suitable for inner content in a squared frame. Fully rounded or square/straight shapes were almost even with 28% and 27% of the answers. From the rounded

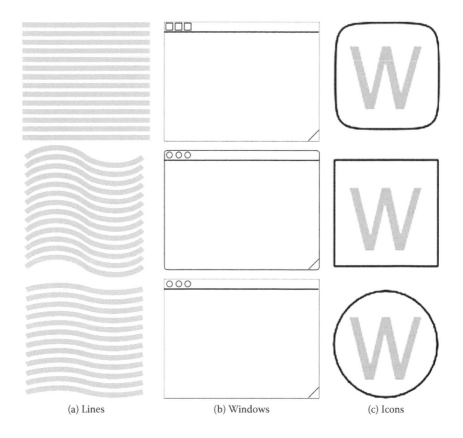

(a) Lines   (b) Windows   (c) Icons

FIGURE 8.33 Shape preferences.

set, windows were the most favorite (59%). Rounded windows were found to have character and be comfortable, kind, and soft. From the square/straight set, dividers led with 81%. Straight dividers were found to be clear, simple, and comfortable. Other shapes tended to feel formal and rigid. (See Table 8.1.)

Czech

Results summary.

The most favorite shape was a radius with 52% of the responses. The majority chose icons (42%), windows came second (32%), and dividers third (26%). Icons

---

TABLE 8.1
Shape Preferences in the Chinese Sample

Square/Straight (16)	Divider 81	Window 19	Icon 0
Rounded (17)	Divider 6	Window 59	Icon 35
Radius (27)	Divider 22	Window 35	Icon 52

TABLE 8.2
Shape Preferences in the Czech Sample

Square/Straight (16)	Divider 60	Window 10	Icon 10
Rounded (13)	Divider 0	Window 40	Icon 25
Radius (31)	Divider 40	Window 50	Icon 65

are sympathetic and organic. Fully rounded or square/straight shapes were almost even with 27% and 22%. From the rounded set, windows were the most favorite (62%). Rounded windows were found to be soft, but still fully intelligible. From the square/straight set, dividers lead with 75%. Straight dividers were found to be clear and to counterbalance the other elements. (See Table 8.2.)

Discussion/conclusion.

Both the hypotheses were fully supported by the results. No significant differences were found between both of the tested groups.

30) If you designed a website, would you make it look similar to a well-known one?

Chinese

Hypothesis.

Copied elements are better perceived than original elements.

Results summary.

Most of the respondents (55% + 5%) would design a website looking similar to well-known ones. The main reasons would be familiarity with the way the content is presented on a layout gained from experience. The users stressed the convenience of such an approach, because it builds on proven design principles, thus saving time in finding the right information. Websites should differ between one another rather through content. (See Figure 8.34(a).)

On the other hand, 35% of the respondents disagreed: In websites they would look for interesting, inspiring, unique, or even breakthrough designs. The reasons for this would be to evolve the current design and to have the websites cater to the user's needs. Also, a preference for simple, uncluttered UI was cited.

Czech

Results summary.

Most of the respondents (55% + 5%) would design a website looking similar to well-known ones. The main reasons were familiarity with the way the content is presented on a layout gained from experience. The users stressed the convenience of such an approach because it builds on proven design principles, but also fear from the unknown. Users would prefer similar websites. On the other hand, 20% of the respondents disagreed: In websites they would look for original, novel designs. Also,

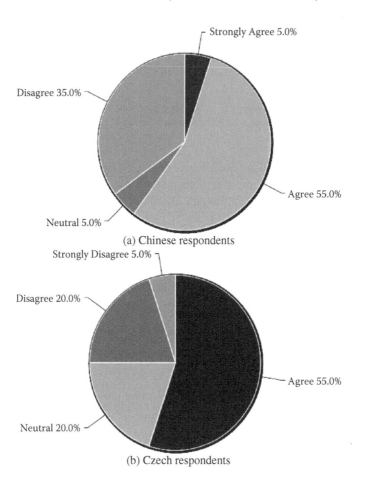

FIGURE 8.34 Perception of copied elements.

20% of the respondents were neutral, and wanted to strike a balance between usability and creative attitude. (See Figure 8.34(b).)

Discussion/conclusion.

The hypothesis was supported by the results. However, what was not clear enough was the degree of similarity the users would welcome the most, that is, if the users would react better at some element taken directly from another UI, or would like the copy on a more general level, such as a UI pattern or layout. Also, the results did not tell anything about the cultural inclination towards the art of copy and copying present in Asia. About a third of the respondents disagreed and supported original and novel designs. Between the Czech and Chinese respondents there were no significant differences, except from the side of those who disagreed. The Chinese results showed a stronger difference between affirmative and disapproving responses, while the Czech results contained more neutral votes.

8.1.3 COLOR

The respondents were presented a 16 color palette Video Graphics Array (VGA) based on the Hyper Text Markup Language (HTML) 4.01 standard (`http://www.w3.org/TR/REC-html40/types.html#h-6.5`). The supported hypotheses regarding colors and color combinations were:

- Users would prefer lighter (pastel) colors and a white background.
- Personal websites would use a wider color palette than websites for other purposes.

Unsupported hypotheses were the following:

- UIs with the white and yellow colors in the foreground tend to be regarded as more aesthetic.
- Background color is more important than foreground color. Interestingly, the Czech sample results supported our hypothesis and valued the background highly.
- UIs with the following background/foreground color combination are most appealing: white on blue, white on gray blue, white on purple. The background color preference was shared among the groups, except for lime, which was chosen by the Chinese. For foreground, blue was a favorite for the Chinese, while red and silver for the Czechs. From the shared color combinations, black on white stood for clearness and naturalness for the Chinese, while for the Czechs it indicated contrast and simplicity.

Moreover, some interesting insights into the perception of colors emerged, as we can see in the following question 31.

31) What features or actions do the following colors imply to you?

Chinese

Results summary.

> Black: Midnight/night/burial/death/dark/depressed/afraid (10), dignified/solemnity/mysterious/serious/stately/quiet/mature (8), blackboard/pencil/ink/thinkpad (5), steady/persist (2), cool (2), winter/cold, bad people, strange, cold blood, hair, angry, suit showing one's temperament, not for websites, deep
>
> Navy: Color for painting/ink/type (5), sky (2), navy (2), depressed/sorrowful (3), old coat/uniform (2), profession, deep, rare, teachers' tables and chairs, desktop wallpaper, uncomfortable, nail polish, reminiscence, quiet, treat, skulduggery, accountant, mature, mysterious, sports shoes, toy, relax
>
> Green: Grass/tree/plant/pine/lime (11), spring/summer (3), environmental protection (2), fresh/active, precious stone, army uniform, old, primary school, traffic light, industry, color blindness test, classmate's favorite, Chrome browser theme, life, coat
>
> Teal: Ink/paint brush (2), bright (2), dark, button, older desktop, immature, unfamiliar, depressed, lake, gem

Silver: Machine/car/watch/metal/ring/appliance/robot (8), shoes/elegant/stage/coat/ fashion/gorgeousness (6), cold/cool/rain (5), nervous, violent, solemn, scroll bar, inactive web-link, radiation, elephant, depressive

Blue: Sky/cloud/ocean/wide/big (12), clean/pure/limpidity (4), sedate (3), smile face/happy (2) gentle, soft, sports, screen, shirt, flag, mature, mysterious, Italy, navy

Lime: Nature/grass (land)/leaves/forest/wood/vigor/spring/young/origin of life/life (17), comfortable/relax (4), children, QQ, waiting pointer index, picture by a child, bright, clear, food package, pen, petting when sleeping, joviality

Aqua: Sky/lake/ocean (6), brightness/bright (2), positive emotion/happiness (2), steady, mature, lime, color in Word, fresh, clean, quietness, apple, weird, wall

Maroon: Hair (5), jujube fruit/chestnut/pine nut (3), blood (2), dark (2), unimportant things, mysterious, unfamiliar, old women coat, firewood, afraid, reminiscence, hotel, flower, mature, bark, nail polish

Purple: Clothes/full dress/bracelet/nail polish/sexy/worldliness/elegance/elegant/fashion (10), flower/grape (5), elegance/elegant/fashion (4), mysterious (2), amethyst, lavender, older fuchsia, website text, romantic, dignity, artistic, quiet, lucky, activity

Olive: Army/army coat (5), dead tree, coconut, plants/nature (3), quietude/quiet (2), soil/earth (2), order, health, simple, old, warm, plain, comfortable, personality, pessimistic, rugger [T-shirt]

Gray: Gloomy/dim/dark/uncomfortable/depressive/sorrowful/cry/somber/pain/negative (16), rainy/overcast sky/dark day (3), dust/pollution/dirty (3), deep, status/ menu bar, mouse, hair, weak, pencil writing, gray zone of law, brick and earth, inactivity

Red: Passion / enthusiasm / vitality / young people / active / energy / blood / good things (13), national flag (5), hot/warm/sun/candle/summer (4), important (3), girl's (coat/dress)/lipstick/nail polish (3), jollity/want to laugh/happiness (3), wedding, sudden, angry, pain, spring festival

Fuchsia: Girls' coat/makeup/apparel/bag/little girl (6), flower (5), lovely/beautiful (4), bright/obvious/light (2), soft/tender (2), cabbage, young (2), naive, warm, chemistry, bear, good mood, wine, activity

Yellow: Bright/obvious/sun/sunlight/warm, optimistic/good mood/excited/ passion, full of energy/activity (14), orange, gold, corn, autumn, oil, lemon, cream, cake/ cream puffs (8), childhood/child/young/naive/fresh (5), traffic light (2), depression, mild, dazzle, moon, paper/cartoon, Brazil

White: Clean/clear/pure (8), wedding (6), naive/origin/holy/simple (5), hospital/nurse/medical/surgical (5), snow (3), cloud (2), wall (2), burial, holidays by the sea, butter, ice cream, website, coat, chalk

Czech

Results summary.

Black: Death/burial/grief/destruction/darkness/night/sleep (16), important/elegance/business/social/solid (7), information/books/font/text/terminal (9), contrasting (2), invisible, black is not a color, coat, shoes, doormat, coal

Navy: Official documents/conservatism/right-wing/international organization (4), navy/sea/wide horizon (4), starry sky/night (3), ink/writing (2), coming of darkness/too dark (2), deepness (2), border, bland, happiness, calm, active relax, strong, dark rainy clouds

Green: Relaxing/calm/harmony/security/balance (11), trees/plants/woods, gardening/lawn/running and sitting in lawn (9), spring mood/summer (2), sleeping/bed linen (2), activism (Greenpeace), do not mind, universal, free, good for the eyes, does not look good on a monitor

Teal: Water/swimming pool/aquapark/sea (5), calm/relax (3), clothing (2), meditative, good on buttons, neutral environments, does not interfere or do something, coral, uncertainty, night sky, end of work, Windows 95, brushing teeth

Silver: Christmas/festivity preparation/tradition/trimmings/winter/snow/snow balling (10), jewellery/exclusive cars/luxury/solid (6), cold/impersonal/technical/stainless steel/metal (6), breaking news, burial with black, modern, background, unusual, wedding portal, cleanness, background for object presentation

Blue: Sea/sky/open/airy/clouds (6), fun/joke (4), neutral (3), cool/cold (2), ball pen/text (2), Facebook, Windows error messages, elegance, calm, positive, clean, talking to kind people, child websites, right-wing party

Lime: Spring/leafs/freshness/grass/health/life/motion/active (8), not for a monitor highlighter/selection/text highlighting (4), food/cafe/restaurant (3), expressive/distinct (2), positive, relax, bite a lemon, provoke, aromatizing, background, too sweet, pistachio, modern

Aqua: Sky/sea/water/swimming/swimming pool (5), angels/innocence (3), vacations/ excursions (2), highlighting (like lime)/attention (2), energy/eagerness (2), signal, skiing, freezing to death, passivity, background, calm color, too radiant, clean

Maroon: Wine/passion/blood/meat/aggressive (7), women/women's apparel (2), too strong for a website/suitable for certain websites (2), chocolate, classical, elegant, luxury, exotic, very strange feeling, digging out potatoes, landscape, comfortable, online game, tiring, serious situation, librarianship

Purple: Special occasions, partying, luxury, trendy, cosmetic products (5), lollipop color/sweets/candy floss (3), roommate's PC, Christmas, well-being, enjoying, blood, hospital, logotype, not very unisex, comfortable, velvet, kitsch, darkness, burglar, bottom, turning, holding, black wizard color

Olive: Dirt/dirty (2), moss/wood wetness/wood smell (3), distasteful color, very negative feeling, carpet, plaster, normality, inconspicuousness, neutral, not for a large surface, well-being, can be interesting, relaxing, reading of an inspirative intelligent book, olives/food/Italy (3), autumn/crop (2)

Gray: Common/neutral/unattractive/boring (6), subways/city/haste/traffic/dirty snow/tubes (6), administration/office (2), modern, elegant, sad, mist of distance, use gray as a last resort, unimportant, solid, background, not current/cancelled information (2), IT/machines (2), complementary (2), quarrel/unpleasant feeling (2)

Red: Active/aggressive/positive/heart/blood/erotic/passion/sex (12), alarming/ alert/warning (6), strawberries/cherries/fruits/flowers (4), cheerful/fun (2), tiring, wide application, sunset, advertisement, ambulance, meat/appetite

Fuchsia: Girlish (3), blogs/photos (2), candies, fashion/elegance (2), rainbow, spoiled, flowers (peony), perfume, warm pleasant color, looking at an awfully dressed girl, last resort for text highlighting, cunning, visited links, terrible pink, cheerful, furiously eroticizing, kitsch

Yellow: Brightness/positive/activity/sun/flowers/warm color/warmth/good mood (16), warning (with black)/attention/highlighting (7), trip to Africa/desert (2), inedible, juice, ideas, beginning

White: Cleanness/clean/snow/cleaning laundry (10), neutral/not interesting/ background/serious/neutral-negative (6), beginning/empty paper/potential (4), untrustworthy places (hospitals, shopping malls, large companies)/hospital/doctor/blind (4), contemplation/meditation (2), calmness, hope, wedding, modern, positive, I like it, can be combined with every color, pain, freedom, emptiness

Discussion/conclusion.

Because we chose for the test a simple color palette with 16 members, orange was omitted. The respondents were asked to choose between yellow or red. Some interesting color associations were discovered that show the given cultural semantic fields. The items are grouped according to the number of occurrences. The sorting of items was done according to shared attributes and/or associations, which were then compared between both of the groups. The resulting groups of semantic items were compared between the Czech and Chinese user groups with the following observations:

- Black: The largest group of semantic items pertained to night and death, followed by solemn and elegant, and ink and information device.
- Navy: The largest shared groups contained painting and writing, sky, navy, and sea.
- Green: The largest shared groups were related to grass and plants, spring and summer, and a commonality was also found in environmental activism.
- Teal: The Chinese respondents connected this color to ink and paint, whereas the Czech respondents connected it with water and swimming as well as relaxation. No shared meaning was found.
- Silver: Metal and machine was the most common shared group of items, followed by fashion and luxury. The Chinese respondents mentioned also cold and rain, while the Czech respondents mentioned Christmas decorations and snow.
- Blue: The most common shared meaning was related to the sky and ocean, followed by happy and fun. The Chinese group mentioned clean and relaxing while the Czech group regarded blue as neutral and cold.
- Lime: Spring, life, and vigor was the largest and single shared meaning of this color. For the Chinese, it was also a color of comfort and relaxation while for the Czechs it was connected with food and eating.

- Aqua: Sky and water was the most common meaning shared between the two groups, the second was positive and energy. For the Chinese group, it was a bright color while for the Czech respondents it was connected with angels and innocence.
- Maroon: Passion and blood was the most common shared meaning. The Chinese respondents also answered hair and chestnuts, while the Czech respondents said women's apparel.
- Purple: Dress, luxury, and attractiveness was the largest and single shared meaning. For the Chinese group, the color was connected also with flowers while the Czech group named sweets and candy floss.
- Olive: No directly shared meaning was found. The Chinese respondents regarded the color as military or soil oriented, while the Czech respondents connected with it wood, food, and dirt.
- Gray: The single shared meaning was gloomy and unpleasant, mostly cited by the Chinese respondents. For them, it was also connected with rain and dust, while for the Czech group the color pertained to neutral, city, and office.
- Red: Together with maroon, the largest shared group revolved around passion, blood, and energy. Alarm and warning came second, followed by happiness and fun. For the Chinese respondents, it was a color of the national flag, warmth, and lipstick while for the Czech group it was connected with berries and flowers.
- Fuchsia: The single largest shared group was connected with girlish fashion and apparel. For the Chinese group, it was also a color of flowers and beauty.
- Yellow: The single largest shared group orbited around brightness, warmth, good mood and activity. For the Czech respondents, it was a color of attention and highlighting while for the Chinese group it was for childhood and freshness.
- White: The largest shared group was related to cleanness, followed by hospitals. For the Czech group, it was connected with neutrality, potential, and contemplation. The Chinese related the color to simplicity and holiness, snow, but also weddings (together with red).

32) Which background color do you like the most?

Chinese

Hypothesis.

UIs with the following background colors tend to be regarded as more aesthetic: blue, purple, cyan, or gray.

Results summary.

The largest group (25%) voted for white color for background. Some users are used to white as the background and they think white can match other colors well, and at the same time, lets the foreground emerge. The second largest group (15%) chose aqua. Users consider aqua to be clear, and it is neither dull nor showy. When users see aqua, they feel comfortable and refreshed. (See Figure 8.35(a).)

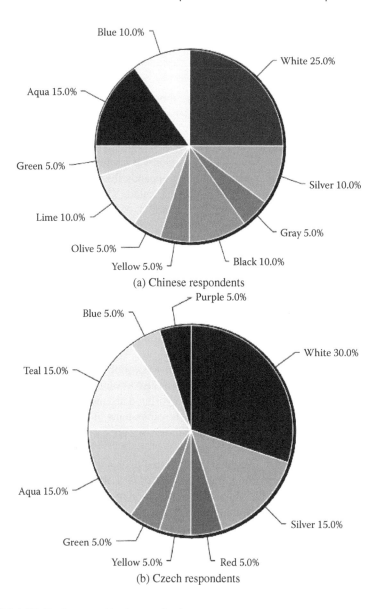

FIGURE 8.35 Preference of background color.

Czech

Results summary.

The largest group (30%) selected white as their choice for a background. The reasons were high contrast with black text, but also good combination with other colors. Aqua, teal, and silver came all second (15%). Aqua was found to be optimistic, comfortable, and well readable; teal and silver were only thought of in combination with other colors. (See Figure 8.35(b).)

Discussion/conclusion.

The hypothesis was partly supported by the results. In the Chinese group, the preference for blue, fuchsia (purple), aqua (cyan), and gray (silver) amounted to 40% of the responses. However, the most frequently cited colors were white and aqua (cyan). In the Czech group, the preference for aqua (cyan), silver, and teal amounted to 45%, but white was chosen by the largest part of the respondents. In both of the groups, white was the most popular, followed by aqua. Every choice was backed by a practical reason. It does not seem to be grounded in a habit.

33) Would you prefer a lighter or darker color (chroma)?

Chinese

Hypothesis.

Users would prefer lighter (pastel) colors, with a white background.

Results summary.

The majority of the respondents (63%) preferred light or lighter colors. Interestingly, many of the answers cited that bright colors evoke happy, positive feelings, which seemed to be very important for the respondents. Brighter colors were not rated too positively, however, because they might hurt the eyes in the respondents' opinions. Dark colors, on the other hand, would seem depressive for these people. Only a minority (26%) would prefer dark or darker colors in the UI. The reason these respondents gave were based on ergonomics, as one can look longer at darker colors and still feel comfortable. (See Figure 8.36(a).)

Czech

Results summary.

The majority of the respondents (60%) preferred light or lighter colors (almost spread evenly between the two). The reasons were less eye strain, and an optimistic outlook. Only 20% chose dark colors, because they provide a better contrast. (See Figure 8.36(b).)

Discussion/conclusion.

The hypothesis was supported by the results. We noticed a strong causal relation between the color chroma and user's mood in the Chinese sample. Much preferred were light colors, as these would inspire life and good feelings in the respondents. In the Czech sample, the reasons were more grounded on habits and practical reasons, but mood was also a strong factor.

34) Which foreground color do you like the most?

Chinese

Hypothesis.

UIs with the following foreground colors tend to be regarded as more aesthetic: white or yellow.

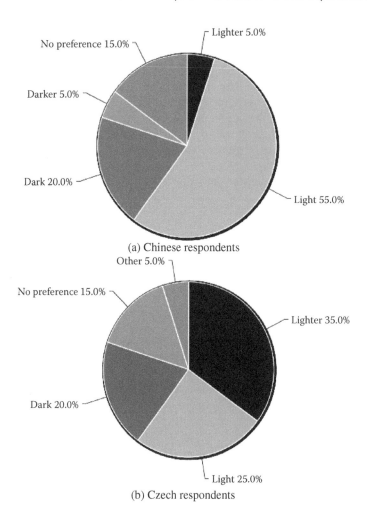

(a) Chinese respondents

(b) Czech respondents

FIGURE 8.36 Preference of color chroma.

Results summary.

The majority of the responses was split in three colors: black (25%), blue (20%), and lime (20%). According to the respondents, black would be clear enough and match other colors well. Blue was chosen not only for practical reasons (good match with other colors), but people expressed positive emotions towards it (they liked it), and they were also used to it from websites they would visit. Lime was found to be light, comfortable, clear, fresh, and to have a good influence on mood. (See Figure 8.37(a).)

Czech

Results summary.

The largest group of the respondents chose black (25%) for their preferred foreground color, followed by white and navy (both 15%). Black makes for a good contrast, and

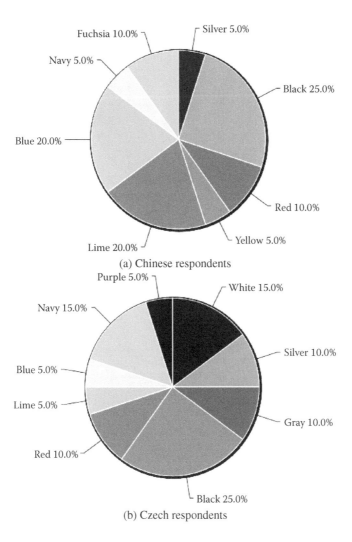

Fuchsia 10.0%
Silver 5.0%
Navy 5.0%
Black 25.0%
Blue 20.0%
Red 10.0%
Lime 20.0%
Yellow 5.0%
(a) Chinese respondents

Purple 5.0%
White 15.0%
Navy 15.0%
Silver 10.0%
Blue 5.0%
Lime 5.0%
Gray 10.0%
Red 10.0%
Black 25.0%
(b) Czech respondents

FIGURE 8.37 Preference of foreground color.

is not irritating. Navy is deep, and serious. White was chosen also for its contrast features on a darker background. (See Figure 8.37(b).)

Discussion/conclusion.

The hypothesis was not supported by the results. Only one Chinese respondent chose yellow for the foreground; white was not cited at all. The results were based on practical, as well as emotional reasons. Interestingly, both the Chinese and Czech respondents shared black, while blue and navy was chosen as second. Lime was chosen as second in the Chinese group, but almost omitted in the Czech group.

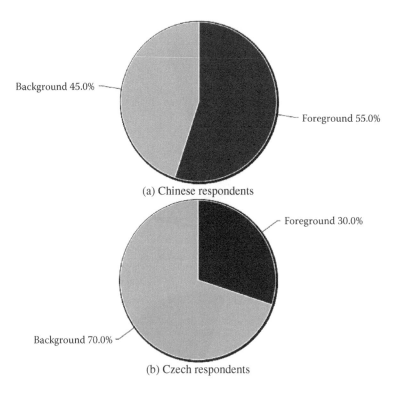

FIGURE 8.38 Preference of color placement.

35) On a screen, which color placement is more important to you?

Chinese

Hypotheses.

- Background color is more important than foreground color.
- Users would notice the UI context (e.g., background) first.

Results summary.

A slight majority of respondents (55%) chose the foreground color over the background color. The prevalent reasoning was that the foreground contains the most important information and it is thus in the center of attention. However, the background color is important in that it sets a general framing regarding the theme of the website or the mood or atmosphere it seeks to convey. (See Figure 8.38(a).)

Czech

Results summary.

A vast majority of the respondents (70%) found the background color placement to be more important. The importance, however, is grounded in the legibility of the foreground information, which an improperly chosen background color can hinder. Also, the background builds the main effect of the webpage. (See Figure 8.38(b).)

Discussion/conclusion.

The first hypothesis was not supported by the results, even if not by a very significant amount. The data from the Chinese sample showed the respondents pay quite a lot of attention to the background for context cues and orientation. However, we could not assess, whether the context (second hypothesis) was noticed first. Interestingly, the Czech sample results supported our hypothesis, and valued the background highly.

36) Which of the following color combinations are most suitable for the following websites?

Chinese

Hypothesis.

Personal websites would use a wider color palette than websites for other purposes.

Results summary.

In all the categories, white was the clear winner, followed by blue and black. According to the most frequent responses, commercial sites would contain white (32, 5%), silver (20%), and black (15%). The color combinations stand for style, formality and serious presentation. Personal sites would contain white (20%), blue (17, 5%), black (17, 5%), and aqua (12, 5%). The color combinations stand for warmth, freshness, cuteness, and happiness (together with lime and fuchsia). Educational sites would contain white (37, 5%), blue (25%), and black (15%). Here, the choices underline clearness, formality, seriousness, and maturity. Also, some respondents mentioned they knew the colors from the Dalian Maritime University (DMU) website. Lastly, governmental sites would contain white (35%), red (17, 5%), and black (12, 5%). The combinations should express stability, authority, seriousness. Red stands for the national flag, the Communist party, and enthusiasm for the masses.

Czech

Results summary.

In all the categories, white was present as the first choice, followed by blue. According to the most frequent responses, commercial sites would contain white (20%), silver (17, 5%), but also blue and black (each with 12, 5%). The color combinations stand for luxury, but also solid, reliable, clean. Personal sites would mostly contain white (20%), and maroon (12, 5%). The combinations stand for warmth, experimentation, and harmony. Educational sites would be based on yellow, white (each with 15%), and blue (12, 5%). They are optimistic, formal, and eye-catching. Governmental sites would contain white (30%), blue, and black (each with 12, 5%). The colors should be formal, accessible, traditional, and reliable.

Discussion/conclusion.

The hypothesis was supported by the results. Although the data show color themes acquired from experience with different types of websites, they also point at a stark contrast between what would be acceptable for formal uses, and what would be suitable for private use. The results were comparable between the Czech and Chinese groups.

37) How do you feel about red background color with yellow text?

Chinese

Hypothesis.

Red color with yellow text is used for special occasions (festivities, family gatherings, national holidays).

Results summary.

The majority of the respondents (65%) reacted negatively to the color combination. The reason for it was low readability, and because it is not widely used. On the Web it might be encountered with spam. A smaller group (25%) reacted positively, because the colors conveyed feelings of warmth and happiness. Only a small group of respondents was neutral (10%). On the whole, the respondents connected the colors with the national flag, the government, banners, and also weddings, hotels, optimism, and clothes.

Czech

Results summary.

Half of the respondents (50%) reacted negatively to the color combination, mostly because of very low readability and usage in spam. A smaller group reacted neutrally (40%). The results were mostly mixed between poor aesthetics and low readability; but on the other hand, such a combination would get a user's attention. Only 10% of the respondents reacted positively, and would use the colors themselves.

Discussion/conclusion.

The hypothesis was partly supported by the results. Interestingly, although many respondents liked the colors, the majority of them looked at the color combination from the perspective of usability. Therefore, the respondents did not like the combination, when it came to practical usage and readability. Interestingly, the Chinese respondents reacted more negatively, than the Czechs, who had more diverse views on the subject.

38) Which of the following color combinations do you like?

Chinese

Hypothesis.

UIs with the following background/foreground color combination are the most appealing: white on blue, white on gray blue, or white on purple.

Results summary.

The most popular background colors were white (25%), lime (20%), and silver (15%). The most popular foreground colors were black (25%), and blue (25%). The most popular combinations were silver on black (10%) standing for simplicity and concentration, blue on lime (10%) standing for relaxation and warmth, and black on white (10%) standing for clarity and naturalness.

Czech

Results summary.

The most popular background colors were white (45%) and silver (15%), while the most popular foreground colors were black (20%), red, and silver (each with 15%). The most popular combinations were blue/navy on white (15%), which are relaxing, harmonic, and allow for other colors to be added. Red on white (clean and warm) and black on white (simple, contrasting) followed (each with 10%).

Discussion/conclusion.

The hypothesis was not supported by the results. This might be due to a different method employed, as we did not show the respondents sets of different color combinations (as in the research the hypothesis is drawing from), but let them combine the colors by themselves. The background color preference was shared among the groups, except for lime, which was chosen by the Chinese. For the foreground, blue was a favorite for the Chinese, while red and silver for the Czechs. From the shared color combinations, black on white stood for clarity and naturalness for the Chinese, while for the Czechs it indicated contrast and simplicity.

8.1.4 SYMBOLS

For testing symbols we used various examples of computer icons found in different applications or we created the examples. The supported hypotheses regarding user's preferences for different types of media and preferences for culture-specific content and trustworthiness of the content were:

- Icons presenting situations are more intuitive than those containing objects. The Czech sample preferred image icons to those presenting situations, in contrast with the Chinese results.
- There is a close similarity between sequential information structure in language and the horizontal structure in visual composition. Verb (a pointer index) and adverb (a "+" sign) would mimic their position in sentence (i.e., the verb comes before the adverb).
- Users can recognize visual patterns occurring in the UI.
- Copied UI elements are better perceived than original elements. This applies both on computer icons and design patterns.
- The sequence of input in a faceted search follows the sequence of natural language. The subject comes first (relating to the user's gender, or size), followed by an implied verb and adverb (purpose) and finally the object (price,

color, rating etc.). In contrast to the Chinese results, the Czech respondents would put size after gender (instead of purpose), purpose instead of price, and price last, thus omitting color and rating.

- The use of Chinese calligraphy was praised by the users.

The unsupported hypotheses:

- Icons presenting images are more intelligible than those containing characters.
- There is a close similarity between sequential information structure in language and horizontal structure in visual composition. Noun (folder) and adjective (star attribute) would mimic their positions in a sentence (i.e., the attribute precedes the subject).
- Long textual pages are considered more useful than texts on more screens because the former contain all the information in one place (show more context).
- Icons with symbols coming from users' own cultural background are better perceived and understood than those from a foreign culture.
- Given that most of the websites contain mostly text, text would be regarded the most useful media. In the Chinese sample, pictorial media (images, videos) had the highest acceptance and credibility. In contrast, the Czech respondents preferred images and texts to videos and sound, both in terms of the efficiency of information transmission and trustworthiness.
- A localized UI would be better accepted than a nonlocalized UI. The respondents were accustomed to using foreign, nonlocalized applications, so localization was their least concern. On the other hand, speed and usability were major concerns among users. Also, originality and aesthetics were highly praised. The Czech results, on the whole, and in contrast to the Chinese results, showed a preference for features instead of color.

39) Please group the following objects (using one line per group).

Chinese

Hypothesis.

Users tend to group objects according to intrinsic relations rather than categories.

Results summary.

The largest group (45%) of the respondents categorized the UI objects by a category (e.g., similarity, areas of presence), followed by almost the same number of responses (40%) categorizing the objects according to the relations between them (e.g., what can manipulate what, what can include what). (See Figure 8.39.)

Czech

Results summary.

The largest group (40%) of the respondents categorized the objects by relation; the others were divided between category, and a blend of the two. (See Figure 8.40.)

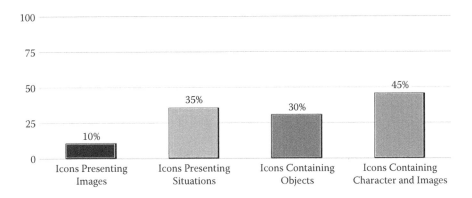

FIGURE 8.39 Preference of different kinds of icons in the Chinese sample.

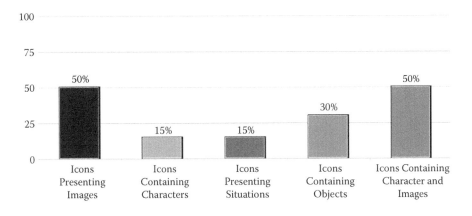

FIGURE 8.40 Preference of different kinds of icons in the Czech sample.

Discussion/conclusion.

The hypothesis cannot be fully verified, because some of the answers cannot be categorized and the difference is not very pronounced. However, the results provide an interesting glimpse into the thinking process involved. In a revision of this test we would reduce the number of objects, and choose objects with bigger differences between them (e.g., pointer, file, folder).

40) Please look at the picture for two seconds and answer the following questions.

This was not asked in the Czech interview.

41) Which kind of icons do you find more useful?

Different examples were shown to the respondents:

• Icons with images (symbols)

- Icons with situations
- Icons with objects

Chinese

Hypotheses.

- Icons presenting situations are more intuitive than those containing objects.
- Icons presenting images are more intelligible than those containing characters.
- Icons presenting objects with a description are more understandable than those without a description.

Results summary.

The most popular icons were those containing characters and images (45%), followed by icons presenting situations (35%). Icons containing objects came third (30%). The main reason for choosing icons with characters and images was their ease of use. When a picture was not intuitive, the characters would help. The icons with scenes were praised for being concise, self-explanatory, and usable even for novice users. Icons with objects were attractive because they depicted their contents well.

Czech

Results summary.

The most popular icons were those containing characters and images (32%), followed by icons presenting images or symbols (29%), and icons presenting objects (19%). Characters with images were found to be the most understandable and accessible for first-time users. Image icons can be faster to use for experienced users, but can be cryptic. Objects share most of the features of images, that is, they must be as simple and as clear as possible to be usable.

Discussion/conclusion.

The first hypothesis was supported by the results. In the Chinese sample, icons presenting situations were regarded as easier to understand and more practical. However, the difference between the two was not very significant (only 5%). The second hypothesis was not supported by the results. Although images (or symbols) were regarded as more attractive than characters, they would often lack a clear meaning for the users. The combination of images and characters obtained the best results in the study, in both the interviewed groups. The Czech sample preferred image icons to those presenting situations, in contrast with the Chinese results. Object icons came third in both of the groups.

42) An icon attribute shows additional features of the object.

Chinese

Hypothesis.

There is a close similarity between sequential information structure in language and horizontal structure in visual composition. Verb (pointer index) and adverb ("+" sign)

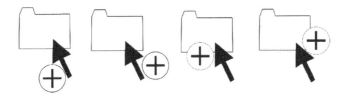

FIGURE 8.41 A plus sign as an icon attribute showing the action of copy.

would mimic their positions in a sentence (i.e., the verb comes before the adverb). (See Figure 8.41.)

Results summary.

The largest group (55%) chose the right positioning of the attribute. Some of the respondents cited habit as their reason along with practicality (the attribute does not interfere with the mouse action) and aesthetics. Three of them understood the plus sign is related to the pointer as an additional information or attribute. Left placement was chosen by the same number of respondents as the below choice (both 15%). The reasons for the left and below placements were all about a possible interference with other objects, given the dragging direction the user is used to. (See Figure 8.42.)

Czech

Results summary.

The largest group (48%) chose the right positioning of the attribute. Left placement came second with 22%, and above came third (17%). The attribute on the right from the pointer follows the habitual placement from operating systems, but also it expands the information in the reading direction. The attribute on the left makes the pointer look more compact, and it refers more directly to the action. The attribute above looks fresh and fun, however. (See Figure 8.43.)

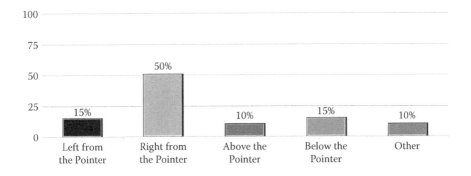

FIGURE 8.42 Preference of the pointer attribute position in the Chinese sample.

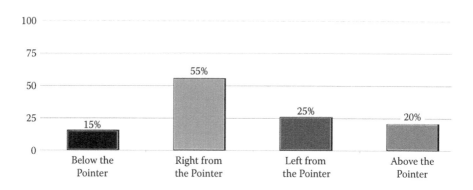

FIGURE 8.43 Preference of the pointer attribute position in the Czech sample.

Discussion/conclusion.

The hypothesis was supported by the results. Moreover, almost a third of the respondents, who chose the right placement, recognized the attribute to expand on the first information (the pointer), in both of the groups. In contrast to the Chinese group, which placed the attribute in the second place below the pointer, the Czech respondents would place it above. However, the result coincided with the habitual placement of the attribute in real life, which may have triggered a familiarity bias. To further verify the hypothesis we should design and perform more tests, using perhaps novel objects for the respondents. We should also investigate further the cases not supporting the hypothesis.

43) An attribute shows additional features of the object.

Chinese

Hypothesis.

There is a close similarity between sequential information structure in language and horizontal structure in visual composition. Noun (folder) and adjective (star attribute) would mimic their position in a sentence (i.e., the attribute precedes the subject). (See Figure 8.44.)

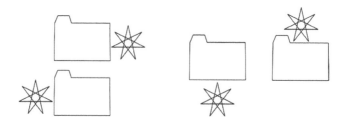

FIGURE 8.44 A star sign as an icon attribute showing the kind of icon.

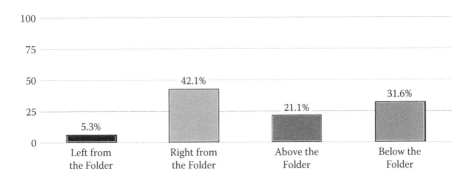

FIGURE 8.45 Preference of the folder attribute position in the Chinese sample.

Results summary.

The largest group (42%) chose to place the attribute right from the icon. Again, the main reason is pragmatic: the placement at right would not interfere with the icon title. Also, the respondents were used to seeing this placement. The below placement came second (32%). Interestingly, here was also the reason of practicality. One respondent was aware such a placement would follow the sequence of attention, where the folder icon comes first. Above the folder placement came third (21%), mainly because it would not interfere with other information and can better catch people's attention. (See Figure 8.45.)

Czech

Results summary.

The largest group (36%) chose to place the attribute above the icon. Second came right placement (23%), and third left placement (18%). The above placement would expand the meaning of the icon, is compact, and follows the reading direction from top to bottom. The right placement was chosen because it is telling something about the icon and was regarded as practical. The left placement would categorize the icons according to their type, and it would be best suitable for a list of items. (See Figure 8.46.)

Discussion/conclusion.

The hypothesis was not supported by the results. However, the tested image may have led the Chinese respondents to focus more on the pragmatics of icon design, rather than on the sequence of elements. Therefore, to further verify the hypothesis we should design and conduct more tests, using perhaps novel or more abstract objects for the respondents, which would not trigger a familiarity bias. We should also investigate further which cases would support the hypothesis. Interestingly, in the Czech sample there were differences in the sequence of priorities, where the above placement was chosen as first, while in the Chinese sample it was only third. Moreover, 14% of the responses would like to see the attribute to be placed in the middle of the icon.

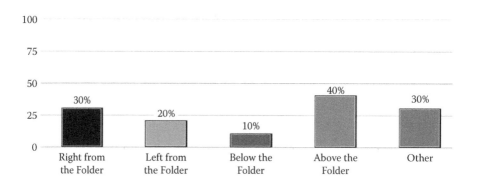

FIGURE 8.46 Preference of the folder attribute position in the Czech sample.

44) On a website, which way of presenting information would you prefer?

Chinese

Hypothesis.

Long textual pages are considered more useful than texts on more screens, because the former contain all the information in one place (show more context).

Results summary.

The majority of the respondents (67%) opted for shorter pages with lower context. Although shorter pages were preferred for information retrieval, they are not so common because of technical limitations of the network (single pages load quicker, even if they are longer). The reasons for choosing shorter pages were that the users do not need to go through much information, do not need to scroll, and do not feel tired or uncomfortable when browsing. Longer pages accounted for a far smaller amount (25%). The reasons for choosing them were that the users have a better overview of the content of the page and the quick loading feels smoother. (See Figure 8.47.)

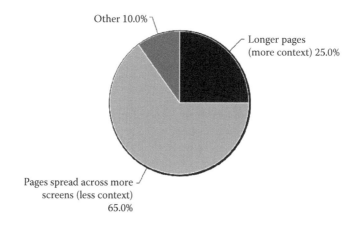

FIGURE 8.47 Preference of textual presentation in the Chinese sample.

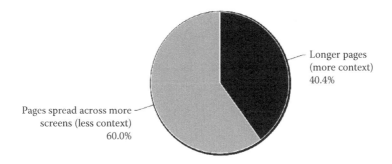

FIGURE 8.48 Preference of textual presentation in the Czech sample.

Czech

Results summary.

The majority of the respondents (60%) chose shorter pages, because they allow for quicker orientation, navigation, and do not force users to read all the information, if they choose, by providing snippets of text. Longer pages were preferred by 40% of the respondents. The reasons were that the users feel more stability and do not need to click, just scroll down. (See Figure 8.48.)

Discussion/conclusion.

The hypothesis was not supported by the results. Although many popular websites contain long texts and few images, the reason is not so much cultural, as it is technical and usability-related. Still, the majority of the respondents would prefer websites with shorter pages, which held true for both of the interviewed groups.

45) Please look at the following icons. What kind of applications would you connect with them?

Chinese

Hypothesis.

Icons with symbols coming from users' own cultural background are better perceived and understood, than those from a foreign culture.

The respondents were presented with Chinese and international applications icons:

- First row: Chinese application icons (C)
- Second row: international application icons (W)

Results summary.

The Western set of icons scored better compared to the Chinese set. The meaning of the movie player icon was recognized by all of the respondents (W3: 100%), followed by the editing tool (W1: 89%), the encryption tool (W4: 84%), and the helmet (W2: 32%). The meaning of the five-input method was recognized by most of the respondents (C1: 95%), followed by the chess game (C4: 74%), the hammer and scythe for communism/government/flag (C2: 53%), and the symbol for house demolishing (C3: only one respondent knew the precise meaning, whereas 37% recognized at least the general meaning of "removal"). According to some of the respondents, Chinese characters could add some spirit to the icons, but might appear unattractive. When used, they should not be too complicated. The foreign icons tend to be more concise and visually appealing.

Czech

The respondents were presented with Czech and international applications icons:

- First row: Czech application icons (C)
- Second row: international application icons (W)

Results summary.

The Western set of icons scored better compared to the Czech set. The meaning of the encryption tool was recognized almost by all of the respondents (W4: 95%), followed by the movie player (W3: 85%), and editing tool (30%). The meaning of the mapping application was recognized by all of the respondents (C3: 100%), followed by public transport ticketing (C1: 90%), and chatting application (C4: 25%).

Discussion/conclusion.

The hypothesis was not supported by the results in either group. The icons for the test were chosen from a Chinese software portal containing applications both from local and foreign developers. The Czech part was taken from a portal of Android applications for the Czech users. For testing the cultural background we chose three icons containing Chinese characters and one with a communist symbol. For the foreign symbols we chose image-only icons: two containing tools (scissors, key) and two containing objects (helmet, movie film). Interestingly, the foreign icon set proved to be more understandable among the respondents. However, apart from the European-style helmet icon (with only 55% responses in the broad meaning space), all of the rest represented items were used in the local culture as well. Therefore, a future test should select visuals with more different semantics.

46) Are you aware of any recurring patterns in the UI?

Chinese

Hypotheses.

- Users can recognize visual patterns occurring in the UI.
- Copied elements are better perceived than original elements.

Results summary.

All of the respondents were able to recognize visual design patterns in the applications they would use. Some of the respondents preferred the similar layouts over different ones. Also, they were able to give examples of similar patterns from websites focusing on social networking, shopping, news portals, search engines, as well as applications for instant messaging and games.

Czech

Results summary.

Almost all of the respondents (90%) were able to recognize UI patterns. Moreover, they gave examples mostly of visual patterns (media players, web browsers, social networks) but also of functions (software installation, mechanics of attraction, double-click for open, etc.).

Discussion/conclusion.

The hypotheses were supported by the results. For a further study, more complex examples could be given to the respondents to let them focus not only on the visual, but also structural and functional planes of a pattern. In fact, only 24% of the Czech responses mentioned action patterns, and only 15% of the Chinese responses.

47) You are looking to buy new clothes at an e-shop. Which sequence would you reorder the search boxes?

Items: Gender, purpose, brand, rating, age, price, color, material, size.

Chinese

Hypothesis.

The sequence of input in a faceted search follows the sequence of natural language.

Results summary.

The majority of the respondents chose gender as the first item in a search facet (65%). The second item was purpose (30%), brand (20%), and gender (15%). The third item was price (25%), color (20%), and material (15%). The fourth item was price (45%), brand (20%), and size (15%). The fifth item was valuation/rating (33%), color (22%), price, and purpose (both with 11%). Although going further the differences between the responses are not more than 5%, we could construct the search facet in the following sequence: gender, purpose, price (/color), price, valuation/rating. (See Figure 8.49.)

Question.

In which direction do the users want to put the search boxes?

Result.

Most of the respondents chose the horizontal direction. They think it will be easy to read and doesn't need too much space to show the detail selection.

Czech

Results summary.

The majority of the respondents chose gender as the first item in a search facet (73%), followed by age (11%), and purpose (11%). The second item was size (28%), purpose (24%), and age (14%). The third item was size (25%), price (20%), color (15%), and

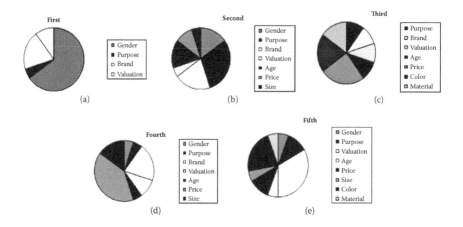

FIGURE 8.49 Preference for the sequence of input facets in the Chinese sample.

age (15%). The fourth and last position answered by all of the respondents was price (23%), size (18%), and purpose (14%). Although going further the differences between the responses are not more than 5%, we could construct the search facet from the following sequence: gender, size, purpose, price. (See Figure 8.50.)

Discussion/conclusion.

The hypothesis was supported by the results. The reasons being, that the subject comes first (relating to the user's gender, or size), followed by an implied verb and adverb (purpose), and finally the object (price, color, rating etc). In contrast to the Chinese

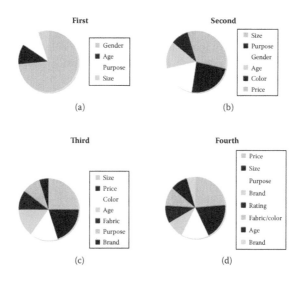

FIGURE 8.50 Preference for the sequence of input facets in the Czech sample.

results, the Czech respondents would put size after gender (instead of purpose), purpose instead of price, and price as the last, omitting color and rating.

48) Please rank the media you find most useful.

Chinese

Hypothesis.

Given that most of the websites contain mostly text, text would be regarded the most useful medium.

Results summary.

The most useful medium was the image (score 83), followed closely by video (81) and text (72). Sound came fourth (44). Also, some of the respondents expressed their opinions about the intuitiveness and credibility of the different media. Pictures in general were the most intuitive and trustworthy, same as videos, which could add more persuasion to the message. Texts, while expressing even more detail, required more attention and were often regarded as untrustworthy. The least popular was sound or voice, because it would not provide more concurrent information, and because it was not very credible.

Czech

Results summary.

The most useful medium was the image (score 78), together with text (78). Video (57) came third, and sound (54) fourth. Images convey information quickly, and are more trustworthy than videos. Text can convey the most information together with details and was found the most trustworthy. Sound comes mostly with video, so it has a similar score of usefulness. Interestingly, intuition (and the "sixth sense") was also cited as a useful medium.

Discussion/conclusion.

The hypothesis was not supported by the results. In the Chinese sample, pictorial media (images, videos) had the highest acceptance and credibility. Texts had still their important role along with pictures, but with much lower trustworthiness. Sound, on the hand, can carry relatively limited information, and often with dubious content. In contrast, the Czech respondents preferred images and texts to videos and sound, both in terms of both the efficiency of information transmission and trustworthiness.

49) Please rank the importance of the following qualities of the UI.

Chinese

Hypothesis.

A localized UI would be better accepted than a nonlocalized UI.

Results summary.

Most of the respondents chose ease of use to be the most important quality of a UI (score 129), followed by speed (119), layout (108), color (106), number of features (93), fun (87), and localization (58). The respondents argued very clearly in favor of useful and usable applications that are visually appealing, fast, and efficient to operate.

Czech

Results summary.

Most of the respondents chose ease of use to be the most important quality of a UI (score 128), followed by layout (118), speed (110), number of features (97), color (79), fun (75), and localization (66). The respondents often understood graphical layout and color to have a direct influence on usability and efficiency.

Discussion/conclusion.

The hypothesis was not supported by the results. The respondents were accustomed to using foreign, unlocalized applications, so localization was their least concern. On the other hand, speed and usability were the major concerns among users. Also, originality and aesthetics were highly praised. The Czech results, on the whole, and in contrast to the Chinese results, showed a preference for features instead of color.

50) Provided two UIs had the same content, would you prefer a UI featuring local artwork?

Chinese

Hypothesis.

The use of Chinese calligraphy was praised by the users.

Results summary.

The majority of the respondents (75%) would prefer calligraphic elements in their UI because Chinese calligraphy is regarded as unique, beautiful, and connected to the tradition. Although it can be used just as an ornament to enhance the UI appeal, it is better to be readable. Few respondents (20%) felt neutrally about the inclusion of calligraphy and questioned the reasons for it. (See Figure 8.51(a).)

Czech

Results summary.

Only a minority of the respondents (30%) would prefer national symbolic elements in the UI. The same number (30%) felt neutral about it, while the largest group (40%) disagreed or strongly disagreed about including such elements in the interface. The reasons were that the respondents did not feel that patriotic to prefer UIs with local symbolism, and would prefer cleaner, leaner applications. (See Figure 8.51(b).)

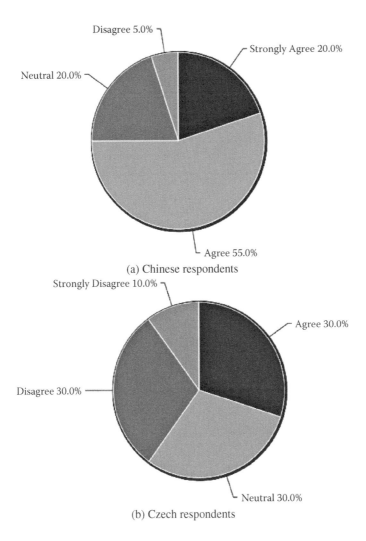

FIGURE 8.51 Preference for local artwork.

Discussion/conclusion.

The hypothesis was supported by the results. Also, the data proved the conclusion of previous research by Kurniawan et al. (2001). Interestingly, although calligraphy was employed at the DMU website, none of the respondents mentioned it. Four respondents (20%) felt very strongly about calligraphy, which resonates vividly in the Chinese culture.

8.1.5 LOOK AND FEEL

For testing the look and feel we used various examples found in different applications or we created the examples. The supported hypotheses in this section regarding user's preference for cartoon imagery, navigation tools, visible and interaction grammar of menus and commands were:

- Menus starting with a verb are considered more natural than those starting with nouns. Although a noun and verb menu was regarded as easy to understand, a verb-driven menu was preferred in that it showed a clear purpose to the user. In contrast, the Czech sample expressed a strong preference towards nouns as these were the most intelligible.
- Cartoon imagery (e.g., little animals) plays an important role in communication. The cartoons improve users' moods and help recall different applications better than characters.

Unsupported hypotheses were:

- Menus progressively disclosing a narrative (e.g., starting with "I want to...") are considered more natural.
- Theme-driven menus (e.g., starting with "I want to...") or role-driven menus (e.g., starting with "I am...") are more logical than menus driven by attributes or concepts.
- The proposed interaction is best understood when starting from a concrete situation (a use case, e.g., "I want to...") rather than user role (e.g., "I am...").

51) Which of the following combinations of menu items do you find more natural?

Chinese

Hypothesis.

Menus starting with a verb are considered more natural than those starting with nouns. (See Figure 8.52.)

Results summary.

The largest group of respondents (40%) preferred verb-only menus. The reasons were, the verb-driven menu expresses the purpose or action the best, and links with nouns to complete the intended action. A combination of nouns and verbs was chosen second (35%), mainly because rigidly following one type of word would limit the possibilities of expression. Third came a menu formed just by nouns (25%). The reason was that nouns better express their meaning, a verb is too general. (See Figure 8.53.)

FIGURE 8.52 Preference for menu types (noun and verb combinations).

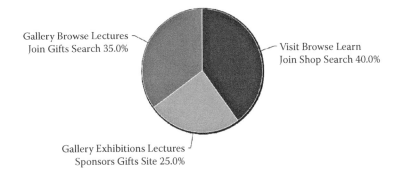

FIGURE 8.53 Preference for menus constructed differently in the Chinese sample.

When asked how the menu would unfold, most users thought the submenu should start with a noun for three main reasons: (1) Nouns represent the object being operated. (2) The main menu should be verb-driven. Verbs show actions. Nouns show targets. (3) The users search in the main menu by verbs. Nouns represent objects.

Some users thought the submenu should be a combination of verb and noun, because it can thus express its purpose more clearly.

Some users think the submenu should start with a verb for three reasons: (1) The main menu is noun-driven. Nouns show a destination. Verbs show actions. (2) The main menu is noun-driven. Nouns are interesting and attractive. Then verbs show functions. (3) Verbs should be clear and complete.

When asked about the direction of the navigation, many users suggested the drop-down list is more natural. Only one user held that the submenu should list horizontally.

Czech

Results summary.

Most of the respondents (55%) chose noun-only menus, because they conveyed their purpose most clearly, and also followed convention. Verb and noun menus followed second (25%, together with respondents-created combinations), and verb-only menus third (15%), although they were found to be clear and more dynamic than nouns. (See Figure 8.54.)

When asked how the menu would unfold, most users thought the submenu should start with a noun, but some users thought the submenu should start with a verb, or a combination of verb and noun. Also, the submenu should complement the main menu to form a sentence.

When asked about the direction of the navigation, users expected the menu to unfold downwards; one expected the menu to jump directly on a new page.

Discussion/conclusion.

The hypothesis was supported by the results. However, in the Chinese sample the noun and verb combination came very closely after (a difference of one respondent).

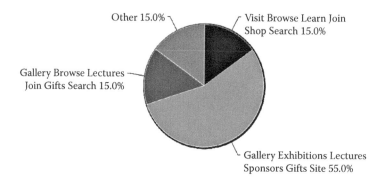

FIGURE 8.54 Preference for menus constructed differently in the Czech sample.

Although the noun and verb menu was regarded as easy to understand, a verb-driven menu was preferred, in that it showed a clear purpose to the user. In contrast, the Czech sample expressed a strong preference towards nouns, as these were the most intelligible.

52) At a university website, where in the menu would you look for course information?

For this question, we used the menu options presented in Figure 8.55.

Chinese

Hypotheses.

- Menus progressively disclosing a narrative are considered more natural.
- Theme-driven menus are more logical than menus driven by attributes or concepts.
- The proposed interaction is best understood when starting from a concrete situation (a use case), rather than a user role.

Results summary.

Most of the respondents chose a concept-driven menu (60%) because it is clearly identified, easy to understand, and fit for a school website, which should be serious and objective. Both the theme-driven ("I want to...") and role-driven ("I am...")

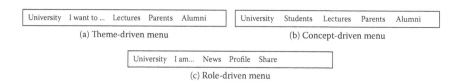

FIGURE 8.55 Preference for menu types (theme, concept, role).

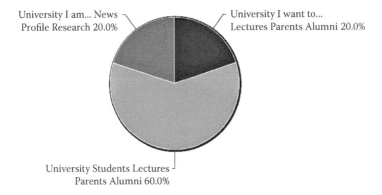

FIGURE 8.56 Orientation in menus constructed differently in the Chinese sample.

menu acquired the same number of responses (20%). The users who chose the theme-driven menu thought they could access all the required operations from the menu, and the menu was more natural and user-friendly. Users preferring the role-driven menu suggested it was closer to life. (See Figure 8.56.)

Czech

Results summary.

The majority of the respondents (70%) favored a concept-driven menu, mostly because it conveyed the navigation structure most directly. The role-driven menu came second (25%), because it was more personable and also compact. (See Figure 8.57.)

Discussion/conclusion.

The hypotheses were not supported by the results. The theme-driven and role-driven menus did not have so much of a following, although the respondents liked them, and found them very natural, because they came from life situations and spoken language. The results were very similar in both of the interviewed groups.

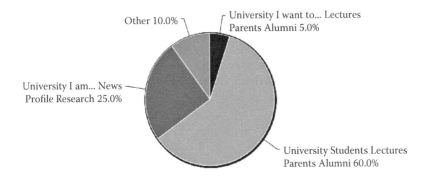

FIGURE 8.57 Orientation in menus constructed differently in the Czech sample.

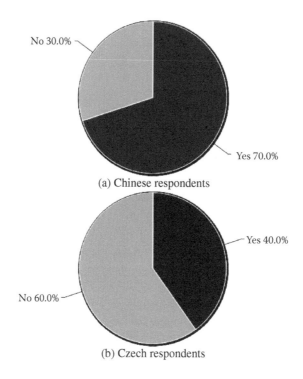

FIGURE 8.58 Preference for cartoon imagery.

53) Do you like cartoon imagery (little animals) present in the UI? Why?

Chinese

Hypothesis.

Cartoon imagery (little animals) plays an important role in communication.

Results summary.

The majority of the respondents (70%) liked cartoons in the UI. The respondents found the UI with cartoon imagery beautiful. In their view, cartoon images would make the users feel relaxed and the website easy to remember. When the network does not load quickly, the small animals may let them not realize how long it takes to load the website. Most of the websites with cartoons are designed for children. On the other hand, those who disliked cartoons suggested they were childish and useless. (See Figure 8.58(a).)

Czech

Results summary.

The majority of the respondents (60%) disliked the cartoons in the UI. Mostly because they were childish, nonfunctional, and get in the way. The rest of the sample liked the cartoons, because they were funny, although sometimes distracting. Their purpose is understood to be to bring pleasure and entertainment and to build an emotional bond with their users; thus, they are more suited for beginners or children. (See Figure 8.58(b).)

Discussion/conclusion.

The hypothesis was supported by the results. We gathered a host of examples the respondents favored, as well as occasions or usage situations (desktop, browser, websites, instant messaging). The cartoons improve users' moods and help recall different applications better than characters.

8.2   PROPOSED GUIDELINES FOR CHINESE UI DESIGN

To help cross-cultural UI designers utilize our findings, we present our results in the form of guidelines that could also be used to enhance the user's acceptance of the UI in a specific culture:

1. Important information should appear in the top-left corner or in the middle center of the screen.
2. New (or problematic) information should appear in the middle center or top center of the screen.
3. Given (or familiar) information should appear in the bottom right or middle right of the screen.
4. Ideal (or general) information should appear in the middle left or top left of the screen.
5. Real (or detailed) information should appear in the middle center or middle left of the screen.
6. Images should be placed in the middle-left or top-right corner of the screen.
7. Put information meant to be easily noticed in the middle-center or top-left corner of the screen.
8. Carefully choose the images. They start the visual narration on the screen, followed by titles.
9. The layout should allow for a central composition (one-column, three-column, central layout).
10. The layout should follow the golden ratio (4:3 or 16:9).
11. Design the layout to be read from left to right. New information should come from the right.
12. Layout dividers should be straight, windows should have rounded corners, and icons should be rounded.
13. UIs should use common patterns so that users can transfer their knowledge from other UIs.
14. Use blue, purple, aqua (cyan), and gray (silver) for the background color.
15. Use light pastel colors on a white background.
16. Use black, blue, and lime for the foreground color.
17. Put more important information on the foreground.
18. For commercial websites, use the combination of white, silver, and black; for personal websites white, blue, black, and aqua. Lime and fuchsia would

also be well received. For educational websites use white, blue, and black. For governmental websites use white, red, and black.

19. Do not put yellow text on red background.
20. Use silver on black, blue on lime, and black on white.
21. Use icons containing characters and images.
22. Place icon attributes on the right from the icon.
23. Create shorter pages with fewer contexts.
24. Search facets should follow the order of the natural language (subject, verb, object).
25. For the highest acceptance and credibility, use pictorial media (images, videos).
26. Above all, the UI should be fast (responsive) and usable as well as aesthetic.
27. When suitable, use Chinese calligraphy elements (readable is better).
28. Form menus from verbs, submenus from nouns. Alternatively, use a combination of verbs and nouns.
29. To improve users' moods and recall, use cartoon imagery in the UI.

9 Discussion and Conclusion

In Part II we showed some interesting and actionable differences between user groups that we can gather from cross-cultural research. Both the groups were exposed to similar computing environments, which led to similar preferences for the UI structure in general. However, we found a few cultural markers that were different and were related mostly to layout and color (for insights into the cross-cultural perception of colors, see pp. 122–123). The impact of the native language grammar on the spatial and logical UI organization was not so pronounced as we expected. Bigger differences between the compared groups came from habits and respondents' cultural background in general.

For the interview and analysis of the layout questions, we used a matrix with three rows and three columns. We chose not to use the matrix used by Kress and Van Leeuwen (2006) (2 × 2 matrix with a center circle), because we needed more granular results. Although the matrix helped us with the analysis by simplifying the data, it might have constrained the respondents by allowing them to choose only a specific field. If we laid the matrix over the collected responses instead, we would get more detailed results.

Given the large scope of the pilot study we were not able to run more tests on topics where the results did not support previous research results, were not clear enough, or promised more interesting data. In a future study, however, we would like to focus more on those areas.

We hope our results and proposed design guidelines will benefit the international HCI/UX design community and will contribute to a discussion on how cross-cultural research could be enhanced by the semiotic and linguistic approach.

The results for each of the analyzed parts are summarized in the interview results analysis section (Chapter 8). Here, we list the following verified hypotheses about Chinese users:

A) Fully supported

Layout.

- Given information (familiar, agreed upon) is expected on the right of the screen.
- A central composition is regarded more aesthetically pleasing than a triptych composition.
- An even number of elements is more preferred than an odd number. Ideal is eight.
- Images placed symmetrically in the middle look better than on the left or right of the screen.

153

- Free-flow layout is easier to use than a grid-based layout.
- Users tend to read from top left towards the center of the screen.
- Left-to-right lines of text are easier to read than top-to-bottom and right-to-left lines.
- There is a close similarity between sequential information structure in language and horizontal structure in visual composition.
- Curves stand for softness (and would be better perceived), while straight lines stand for hardness.
- Rounded corners (curvilinear patterns) are better perceived than square corners (geometrical patterns).
- Copied elements are better perceived than original elements.
- Icons presenting objects with a description are more understandable than those without a description.

Color.

- Users would prefer lighter (pastel) colors and a white background.
- Personal websites would use a wider color palette than websites for other purposes.

Symbol.

- Icons presenting situations are more intuitive than those containing objects. The Czech sample preferred image icons to those presenting situations, in contrast with the Chinese results.
- There is a close similarity between sequential information structure in language and horizontal structure in visual composition. Verb (pointer index) and adverb ("+" sign) would mimic their position in a sentence (i.e., the verb comes before the adverb).
- Users can recognize visual patterns occurring in the UI.
- Copied elements are better perceived than original elements.
- The sequence of input in a faceted search follows the sequence of natural language. The subject comes first (relating to the user's gender, or size), followed by an implied verb and adverb (purpose), and finally the object (price, color, rating etc). In contrast to the Chinese results, the Czech respondents would put size after gender (instead of purpose), purpose instead of price, and price as the last, omitting thus color and rating.
- The use of Chinese calligraphy was praised by the users.

Look and feel.

- Menus starting with verbs are considered more natural than those starting with nouns. Although the noun and verb menu was regarded as easy to understand, a verb-driven menu was preferred in that it showed a clear purpose to the user. In contrast, the Czech sample expressed a strong preference towards nouns, as these were the most intelligible.

- Cartoon imagery (little animals) plays an important role in communication. The cartoons improve users' moods and help recall different applications better than characters.

B) Partly supported

Layout.

- Users tend to attribute more importance on elements placed in the center of the screen.
- Users tend to read from top left towards the center of the screen.
- New information is most readily noticeable in the top-left corner of the screen.
- New information (key, unknown) is expected on the left.
- Images placed symmetrically in the middle look better than on the left or right of the screen.
- Users tend to read from top left towards the center of the screen.
- Images placed symmetrically in the middle look better than on the left or right of the screen.
- There is a close similarity between sequential information structure in language and horizontal structure in visual composition.
- Users tend to read from top left towards the center of the screen.
- New information is more readily noticeable in the top-left corner of the screen.
- The context of information presented is more salient than the information itself.
- Users would notice first the image, then titles, then the body of text.
- Users tend to read from top left towards the center of the screen.
- Left-to-right lines of text are easier to read than top-to-bottom and right-to-left lines.
- There is a close similarity between sequential information structure in language and horizontal structure in visual composition.

Color.

- UIs with the following background colors tend to be regarded as more aesthetic: blue, purple, cyan, and gray.
- Red color with yellow text is used for special occasions (festivities, family gatherings, national holidays).

C) Not supported

Layout.

- Real information (details, concrete image) is expected on the bottom of the screen. The hypothesis was not supported by the results. The majority of Chinese respondents put real information in the middle level of the screen (middle row in the matrix), overlaying it partly on the new and ideal information.
- Square and double-square layout would be more preferred because it is widely used in Asia (the symbol for Earth, Japanese buildings). Instead, respondents preferred a golden-section layout, such as 16:9 or 4:3.

Color.

- UIs with the following foreground colors tend to be regarded as more aesthetic: white and yellow.
- Background color is more important than foreground color. Interestingly, the Czech sample results supported our hypothesis and valued the background very highly.
- UIs with the following background/foreground color combination are most appealing: white on blue, white on gray blue, and white on purple. The background color preference was shared among the groups, except for lime, which was chosen by the Chinese. For foreground, blue was a favorite for the Chinese, while red and silver were for the Czechs. From the shared color combinations, black on white stood for clarity and naturalness for the Chinese, while for the Czechs it indicated contrast and simplicity.

Symbol.

- Icons presenting images are more intelligible than those containing characters.
- There is a close similarity between sequential information structure in language and horizontal structure in visual composition. Noun (folder) and adjective (star attribute) would mimic their positions in a sentence (i.e., the attribute precedes the subject).
- Long textual pages are considered more useful than texts on more screens, because the former contain all the information in one place (show more context).
- Icons with symbols coming from users' own cultural background are better perceived and understood than those from foreign culture.
- Given that most of the websites contain mostly text, text would be regarded the most useful medium. In the Chinese sample, pictorial media (images, videos) had the highest acceptance and credibility. In contrast, the Czech respondents preferred images and texts to videos and sound, in terms of both the efficiency of information transmission and trustworthiness.

- A localized UI would be better accepted than a nonlocalized UI. The respondents were accustomed to using foreign, unlocalized applications, so localization was their least concern. On the other hand, speed and usability were the major concerns among users. Also, originality and aesthetics were highly praised. The Czech results, on the whole, and in contrast to the Chinese results, showed a preference for features instead of color.

Look and feel.

- Menus progressively disclosing a narrative are considered more natural.
- Theme-driven menus are more logical than menus driven by attributes or concepts.
- The proposed interaction is best understood when starting from a concrete situation (a use case), rather than a user role.

D) Not verifiable

Color.

- Users would notice the UI context (e.g., background) first.

Look and feel.

- Users tend to group objects according to intrinsic relations rather than categories.

Conclusion and Future Work

The detailed results of our case studies are presented and discussed in each part. For discussion of the outcome from "Semiotics of Interaction", please see Chapter 5, and for "Culture of Interaction", please see Chapter 9. We appraise our leading theses as follows:

The UI is a means of sharing and interpreting information between systems.

Our thoughts and actions are guided by intrinsic logic rules, supported by the system of language and culture. Language provides an architecture of the design space of HCI/UX. Linguistics and semiotics provide effective methods to solve problems in communication and interaction design. These methods help define the users in their culture, rather than as culture-independent agents. Moreover, each UI stands on a certain paradigm of use which is not always apparent. The UI ideology defines what relations between users and objects can (or should) be made.

The first thesis has been addressed in Part I, "Semiotics of Interaction." We provided an analysis of the UI languages affecting the user's interaction, and presented an evaluation method based on natural and UI language. Our results prove that taking the linguistic perspective in analyzing UIs provides important insights into the way interactive and communicative systems work.

In discussing the HCI ideology (Section 2.3) we presented semiotics as an analytic method especially in its most complex dimension—pragmatics. Pragmatics stands in the design process at the beginning because it forms the strategy and purpose of the developed UI. In the sign context, pragmatics leads the meaning interpretation—what semantics will be assigned to which syntax elements. Not only is this a process of interpretation, but also the whole UI development strategy is subject to HCI ideology to a large extent. Such HCI ideology acquires its specific form in the UI. For the purposes of developing new UIs and also for interacting with the UIs already in place, it is important to know the ways in which pragmatics, as an interpreting principle, is coded and mediated. We can then counter the ideologies by proper education and analysis.

In the UI corpus (Section 3.3) we presented the transcript of interaction sentences forming language games that served together with the actual UI as a basis for HE evaluation and SA analysis. Moreover, the transcript served as input for defining the different elements involved in the interaction language.

The SA provided the expected kind of data (e.g., conventions, connotations, combinations), that gathered a wider context than those from HE. That said, SA can be used to complement the widely used expert evaluation methods, but could possibly be defined to have a higher overlap with HE. In the latter case, SA would need to be evaluated hand in hand with the interaction sentences.

In summary, our study demonstrated the depth of investigation and breadth of insight that SA can achieve in HCI and how this could enhance the current UX practice. Both methods could be merged to provide a best-of-both solution.

We presented different HCI ideologies leading the user interaction. We proposed a framework for analyzing ideology in the context of HCI. The ideology analysis stands on the semiotic foundation presented in Chapter 2 and proved to bring useful insights.

Every sign in HCI is cultural and therefore informational.

The UI provides a lens for reading and writing cultural data. The user's native language and culture determine his or her mentality, rationality, and the discourse involved. By expressing in different systems of meaning (e.g., languages, UIs), we accent different objects and experiences, which results in different insights into the world we live in. When UIs take into account those differences, they can promote both usability and cultural diversity.

The second thesis has been addressed in Part II, "Culture of Interaction." In that part we showed some interesting and actionable differences between user groups, if any, that we could gather from cross-cultural research. Both the groups were exposed to similar computing environments that led to similar preferences for the UI structure in general. However, we found a few cultural markers that were different and were related mostly to layout and color. The impact of the native language grammar on the spatial and logical UI organization was not so profound as we expected. More differences came from habits and cultural background.

The comparison of influences of different cultural backgrounds, namely Chinese and European, tested the evaluation framework based on semiotics, ideology (mental models), and culture. Our framework proved to be valuable in finding and understanding the cultural differences, and gave us insights on how to design for them.

Future work generated by this project includes the following:

1. Develop a pattern visualization for UI language elements. Such visualization would help to quickly compare interaction structures with different kinds of UI in one culture or between two or more cultures.
2. Compare the SA with other semiotic methods. Although we built our method with the knowledge of other semiotic methods, such a comparison would show the specific benefits and/or deficiencies of SA.
3. Extend the SA method (Appendix B) to focus more on the pragmatic and rhetoric function of the UI to allow for a thorough analysis of the inherent HCI ideologies. It would be especially useful to integrate also an appreciation of the cultural background.
4. Extend the SA cross-cultural method to study also different cultural (e.g., rural vs. urban, male vs. female, young vs. elderly) and linguistic environments (e.g., different sentence structure, different writing direction).

Part III

Appendices

Appendix A: Heuristic Evaluation

AM+A uses UI design heuristics adapted from various sources including the following:

- Apple Computer, Inc. (1992). *Macintosh human interface guidelines.* Reading, Massachusetts, United States, Addison-Wesley Publishing Company.
- Marcus, A. (1992). *Graphic design for electronic documents and user interfaces.* Reading, Massachusetts, United States, ACM Press and Addison-Wesley.
- Nielsen, J. (1994). *Usability inspection methods: Conference companion on human factors in computing systems.* Boston, Massachusetts, United States, ACM Press.
- Tognazzini, B. (1992). *TOG on interface.* Reading, Massachusetts, United States, Addison-Wesley Longman Publishing Co., Inc.
- Tufte, E. (1990). *Envisioning information.* New Haven, Connecticut, United States, Graphics Press.
- Tufte, E. (1983). *The visual display of quantitative information.* New Haven, Connecticut, United States, Graphics Press.
- Tufte, E. (1997). *Visual explanations.* New Haven, Connecticut, United States, Graphics Press.

Aesthetic integrity and minimalist design.

Dialogues should not contain information that is irrelevant or rarely needed. Every extra unit of information in a dialogue competes with the relevant units of information and diminishes their relative visibility. Information should be well organized and consistent with principles of visual design.

Consistency and standards.

Users should not have to wonder whether different words, situations, or actions mean the same thing. Follow platform conventions.

Direct manipulation/see and point.

Users should be able to see on the screen what they're doing and should be able to point at what they see. This forms a paradigm of noun (object) then verb (action). When the user performs operations on the object, the impact of those operations on the object is immediately visible.

Error prevention.

Even better than good error messages is a careful design that prevents a problem from occurring in the first place.

Feedback/visible system status.

The system should always keep users informed about what is going on through appropriate feedback within reasonable time.

Fitt's Law.

The time to acquire a target is a function of the distance to and size of the target.

Flexibility and efficiency of use.

Accelerators—unseen by the novice user—may often speed up the interaction for the expert user such that the system can cater to both inexperienced and experienced users. Allow users to tailor frequent actions.

Help and documentation.

Even though it is better if the system can be used without documentation, it may be necessary to provide help and documentation. Any such information should be easy to search, be focused on the user's task, list concrete steps to be carried out, and be concise.

Help users recognize, diagnose, and recover from errors.

Error messages should be expressed in plain language (no codes), precisely indicate the problem, and constructively suggest a solution.

Information legibility/density.

Maximize the amount of data to the amount of ink or pixels used. Eliminate any decorations on charts and graphs that do not actually convey information, such as three-dimensional embellishments. "Less is more" is the rule in information design as every pixel used that does not contribute to information dilutes it.

Match between system and the real world.

The system should speak the users' language, with words, phrases and concepts familiar to the users, rather than system-oriented terms. Follow real-world conventions, making information appear in a natural and logical order.

Modelessness.

For the most part, try to create modeless features that allow people to do whatever they want when they want to in your application. Avoid using modes in your application because a mode typically restricts the operations that the user can perform. Modelessness gives the user more control over what he or she can do and allows the user to maintain context of the work.

Perceived stability.

In order to cope with the new level of complexity that computers introduce, people need stable reference points. To give users a conceptual sense of stability, the interface provides a clear finite set of objects with a clear, finite set of actions. When particular actions are unavailable, they are not eliminated, but are dimmed.

Recognition rather than recall.

Make objects, actions, and options visible. The user should not have to remember information from one part of the dialogue to another. Instructions for use of the system should be visible or easily retrievable whenever appropriate.

User control and freedom.

Allow the user, not the computer to initiate and control actions. Users often choose system functions by mistake and will need a clearly marked "emergency exit" to leave the unwanted state without having to go through an extended dialogue. Support undo and redo functions.

Visible interfaces/WYSIWYG.

Don't hide features in your application by using abstract commands. People should be able to see what they need when they need it. Most users cannot and will not build elaborate mental maps and will become lost or tired if expected to do so.

Severity ratings.

The severity of a usability problem is a combination of three factors:

1. The frequency with which the problem occurs: Is it common or rare?
2. The impact of the problem if it occurs: Will it be easy or difficult for the users to overcome?
3. The persistence of the problem: Is it a one-time problem that users can overcome once they know about it or will users repeatedly be bothered by the problem?

Finally, of course, one needs to assess the market impact of the problem since certain usability problems can have a devastating effect on the popularity of a product, even if they are "objectively" quite easy to overcome. Even though severity has several components, it is common to combine all aspects of severity in a single severity rating as an overall assessment of each usability problem in order to facilitate prioritizing and decision making.

The severity ratings used in this report are described below:

- Severity level 1: Cosmetic problem only—need not be fixed unless extra time is available on the project.
- Severity level 2: Minor usability problem—could impair users' productivity and ability to learn.

- Severity level 3: Major usability problem—important to fix, so it should be given high priority; impacts users' productivity and increases likelihood of errors.
- Severity level 4: Usability catastrophe—imperative to fix this before the product can be released.

Appendix B: Semiotic Analysis

This appendix lists the criteria used in the SA analysis.

Actors, audiences, paradigms.

Define the UI actors (designers, systems/computing agents, and users), who are the intended audiences of the UI, and the leading interaction paradigm that shapes the interaction and communication. Communication from the UI to the actors should be concise, clear, and unambiguous. The audiences can be revealed, for example, by comparing UIs within similar semantic spaces (e.g., functions). The paradigm can be analyzed by the affordances and limits it sets to the UI users.

Symbols.

Different kinds of symbols connote different semantic spaces, cultural backgrounds and address different audiences. The symbols should be intelligible for the audience and should not carry ambiguous, pejorative, or conflicting meaning. Symbols should be chosen to support the structure of the UI language components.

Syntax.

Signs should be used in a minimalistic manner, possibly only once in a given context, and should not be in conflict with context. Similar signs should be placed in similar places to help build a UI hierarchy and user's expectations. The signs should be divisible into identifiable elements and allow for building meaningful chains. The signs should be consistent both internally (within a UI) and externally (across multiple UIs). The system processes are communicated by UI language components, which should heed the expected syntax.

Rhetorical tropes.

The most common rhetorical tropes in the UI are devices of substitution: metaphor, metonymy, prosopopoeia, and synecdoche. The rhetorical tropes used should be intelligible for the audiences and used effectively. The general metaphor of the UI should help users build the correct expectations of future interaction through the use of consistent mental models. The rhetorical tropes often take the form of implied actions.

Interaction phases.

The beginning of the interaction should be consistent with both middle and end. All the parts of the interaction should follow user's expectations and should pertain clearly to the current interaction game. The user should not be forced to perform a different action than he or she intended. The signs present in a UI should lead the user in a sequence towards the goal of the interaction game through a control of narration.

Patterns.

All of the UI language components form different kinds of patterns given their occurrence in different relations (interdependency, determination, and constellation). The number of the interaction sentences should be as low as possible, possibly in the 7 ± 2 range. The interaction sentence should have as few elements as possible. The HCI should form meaningful temporal units. Similar actions should take a similar time to execute or to receive feedback from the system.

Appendix C: Hypotheses

H1: Users tend to attribute more importance on elements placed in the center of the screen.

H2: Users tend to read from top left towards the center of the screen.

H3: New information is most readily noticeable in the top-left corner of the screen.

H4: New information (key, unknown) is expected on the left of the screen.

H5: Given information (familiar, agreed upon) is expected on the right of the screen.

H6: Ideal information (symbolic, general) is expected on top of the screen.

H7: Real information (details, concrete image) is expected on the bottom of the screen.

H8: Images placed symmetrically in the middle look better than on the left or right of the screen.

H9: There is a close similarity between sequential information structure in language and horizontal structure in visual composition.

H10: The context of information presented is more salient than the information itself.

H11: A central composition is regarded more aesthetically pleasing than a triptych composition.

H12: Users would notice first the image, then titles, then the body of text.

H13: An even number of elements is more preferred than an odd number. Ideal is eight.

H14: A free-flow layout is easier to use than a grid-based layout.

H15: A square and double-square layout would be more preferred because it is widely used in Asia (symbol of Earth, Japanese buildings).

H16: Left-to-right lines of text are easier to read than top-to-bottom and right-to-left lines.

H17: Curves stand for softness, while straight lines stand for hardness.

H18: Rounded corners are better perceived than square corners (curvilinear patterns vs. geometrical patterns).

H19: UIs with the following background colors tend to be regarded as more aesthetic: blue, purple, cyan, and gray.

H20: Users would prefer lighter (pastel) colors and a white background.

H21: UIs with the following foreground colors tend to be regarded as more aesthetic: white and yellow.

H22: Background color is more important than foreground color.

H23: Users would notice the UI context (e.g., background) first.

H24: Personal websites would use a wider color palette than websites for other purposes.

H25: Red color with yellow text is used for special occasions (festivities, family gatherings, national holidays).

H26: UIs with the following background/foreground color combination are most appealing: white on blue, white on gray blue, and white on purple.

H27: Users tend to group objects according to intrinsic relations rather than categories.

H28: Icons presenting situations are more intuitive than those containing objects.

H29: Icons presenting images are more intelligible than those containing characters.

H30: Icons presenting objects with a description are more understandable than those without a description.

H31: There is a close similarity between sequential information structure in language and horizontal structure in visual composition. Verb (pointer index) and adverb ("+" sign) would mimic their positions in a sentence (i.e., the verb comes before the adverb).

H32: There is a close similarity between sequential information structure in language and horizontal structure in visual composition. Noun (folder) and adjective (star attribute) would mimic their positions in a sentence (i.e., the attribute precedes the subject).

H33: Long textual pages are considered more useful than texts on more screens, because the former contain all the information in one place (show more context).

H34: Icons with symbols coming from users' own cultural background are better perceived and understood, than those from a foreign culture.

H35: Users can recognize visual patterns occurring in the UI.

H36: Copied elements are better perceived than original elements.

H37: The sequence of input in a faceted search would follow the sequence of natural language. (The user's mother tongue determines his or her expectations of the interaction.)

H38: Given that most of the websites contain mostly text, text would be considered the most useful medium.

H39: A localized UI would be better accepted than a nonlocalized UI.

H40: The use of Chinese calligraphy was praised by the users.

H41: Menus starting with a verb are considered more natural than those starting with nouns. (Assigning objects to actions feels more logical than the other way around.)

H42: Menus progressively disclosing a narrative are considered more natural.

H43: Theme-driven menus are more logical than menus driven by attributes or concepts.

H44: The proposed interaction is best understood when starting from a concrete situation (a use case), rather than a user role.

H45: Cartoon imagery (little animals) plays an important role in communication.

References

ADOBE SYSTEMS. *What's new in Adobe Photoshop CS2* [online]. San Jose, CA: Adobe Systems Incorporated, ©2005. [cit. 2009-09-04]. Available from: `http://www.adobe.com/aboutadobe/pressroom/pressmaterials/creativesuite2/pdfs/ps/PSCS2-WN.pdf`.

ADOBE SYSTEMS. *Adobe®Photoshop®CS2* [software]. Version 9.0.2. San Jose, CA: Adobe Systems Incorporated, ©1990–2005.

ALEXANDER, Christopher, ISHIKAWA, Sara and SILVERSTEIN, Murray. *A pattern language: Towns, buildings, construction* (Center for Environmental Structure Series). New York: Oxford University Press. 1978. ISBN: 9780195024029.

ANDERSEN, Peter Bøgh. *A theory of computer semiotics*. Cambridge, UK: Cambridge University Press, 1997. ISBN: 0521448689.

ANDERSEN, Peter Bøgh. What semiotics can and cannot do for HCI. In: *Knowledge-Based Systems: Semiotic Approaches to User Interface Design*, 2001, 14.8: 419–424. DOI: 10.1016/S0950-7051(01)00134-4.

ANDRE, Terence S. *Determining the effectiveness of the usability problem inspector: A theory-based model and tool for finding usability problems*. AIR FORCE INST OF TECH WRIGHT-PATTERSON AFB OH, 2000. PhD. Thesis. pp. 1–286. Available from: `http://www.dtic.mil/cgi-bin/GetTRDoc?Location=U2\&doc=GetTRDoc.pdf\&AD=ADA378838`.

APPLE. *Macintosh human interface guidelines* [online]. [cit. 2009-09-23]. Cupertino: Apple Inc, ©1992, 2001–2003, 2012. Available from: `http://developer.apple.com/mac/library/documentation/UserExperience/Conceptual/AppleHIGuidelines`.

APPLE. *Mac OS®X* [software]. [Version 10.5.8]. Cupertino: Apple Inc, ©1983–2009.

ASOKAN, Ashwini and CAGAN, Jonathan. Defining cultural identities using grammars: An exploration of cultural languages to create meaningful experiences. In: *Proceedings of the 2005 Conference on Designing for User eXperience*. AIGA: American Institute of Graphic Arts, 2005. p. 35. ISBN 159593250X.

AUSTIN, John Langshaw. *How to do things with words: The William James lectures delivered at Harvard University in 1955*. [Edited by James O. Urmson and Marina Sbisà.] Cambridge, MA: Harvard University Press, 1962, 166 pp. ISBN 9780674411500.

BADRE, Albert. *The effects of cross cultural interface design orientation on World Wide Web user performance*. 2001. [cit. 2011-12-09]. Available from: `https://smartech.gatech.edu/bitstream/handle/1853/3315/01-03.pdf?sequence=1`.

BARBER, Wendy and BADRE, Albert. Culturability: The merging of culture and usability. In: *Proceedings of the 4th Conference on Human Factors and the Web*. 1998, pp. 1–14. Available from: `http://research.microsoft.com/en-us/um/people/marycz/hfweb98/barber/`.

BARNES, Jonathan. *The complete works of Aristotle*. The Revised Oxford Translation. 2 vols. (Bollingen Series, 71: 2). Princeton, NJ: Princeton University Press, 1984, xi, 2487 pp. ISBN 0691099502.

BARTHES, Roland. *Elements of semiology*. New York: Farrar, Straus and Giroux, 1977, 120 pp. ISBN 9780374521462.

BAUMAN, Zygmunt. *Liquid modernity*. Cambridge, UK: Polity Press, 2000, vi, 228 pp. ISBN 0745624103.

BATESON, Gregory. *Mind and nature: A necessary unity*. Cresskill, NJ: Hampton Press, 2002, xviii, 220 pp. ISBN 1572734345.

BERTIN, Jacques. *Semiology of graphics: Diagrams, networks, maps*. Redlands, NY: ESRI Press, 2011, xv, 438 pp. ISBN 9781589482616.

BOGOST, Ian. *Persuasive games: The expressive power of videogames*. Cambridge, MA: MIT Press, 2007, xii, 450 pp. ISBN 9780262026147.

BRANDT, Per Aage. Meaning and the machine: Toward a semiotics of interaction. In: ANDERSEN, Peter Bøgh, HOLMQVIST, Berit, JENSEN, Jens F., eds. *The computer as medium*. Cambridge, UK: Cambridge University Press, 1993, vii, 495 pp., pp. 128–140. ISBN 9780521419956.

BREJCHA, Jan and MARCUS, Aaron. Semiotics of interaction: Towards a UI alphabet. In: KUROSU, M., ed. *Human–computer interaction*, Part I, HCII 2013, LNCS 8004. Heidelberg: Springer, 2013, pp. 13–21.

BREJCHA, Jan et al. Cross-cultural comparison of UI components preference between Chinese and Czech users. In: RAU, P.L.P., ed. *CCD/HCII 2013*, Part I, LNCS 8023. Heidelberg: Springer, 2013, pp. 357–365.

BREJCHA, Jan. Ideologies in HCI: A semiotic perspective. In: MARCUS, A., ed. *Design, user experience, and usability. Theories, methods, and tools for designing the user experience.*, Part I, HCII 2014, LNCS 8517. Switzerland: Springer, 2014, pp. 45–54.

BUCHLER, Justus. *Philosophical writings of Peirce*. New York: Dover Publications, 1955, xvi, 386 pp. ISBN 0486202178.

BUSH, Vannevar. As we may think. In: *The Atlantic Online*. 1945. [cit. 2009-10-08]. ISSN 1072-7825. Available from: http://www.theatlantic.com/doc/194507/bush.

CAPITAL COMMUNITY COLLEGE FOUNDATION. *Adverbs* [online]. 2005. [cit. 2009-11-12]. Available from: http://grammar.ccc.commnet.edu/grammar/adverbs.htm.

CARROLL, John M. *Making use: Scenario-based design of human-computer interactions*. Cambridge, MA: MIT Press, 2000, xiv, 368 pp. ISBN 0262032791.

ASSIRER, Ernst A. Structuralism in modern linguistics. *Word-Journal of the International Linguistic Association*, 1945, 1.2: 99–120. ISSN 0043-7956.

CHANDLER, Daniel. *Semiotics for beginners* [online]. 2001. [cit. 2009-09-20]. Available from: http://www.aber.ac.uk/media/Documents/S4B/.

CHANG, Emily. *Design 2.0: Minimalism, transparency, and you* [online]. 2006. [cit. 2009-05-01]. Available from: http://www.emilychang.com/go/weblog/comments/design-20-minimalism-transparency-and-you/.

CHOONG, Yee-Yin and SALVENDY, Gavriel. Design of icons for use by Chinese in mainland China. In: *Interacting with computers*, 1998, 9.4: 417–430. DOI: http://dx.doi.org/10.1016/S0953-5438(97)00026-X.

CHUN, Wendy Hui Kyong. On software, or the persistence of visual knowledge. In: *Grey Room*, 2004, pp. 26–51. ©2005 Grey Room, Inc. and Massachusetts Institute of Technology. ISSN 1526-3819. Available from: http://art310-f12-hoy.wikispaces.umb.edu/file/view/Software+or+Persistence+of+Visual+Knowledge.pdf.

CLEMMENSEN, Torkil and ROESE, Kerstin. An overview of a decade of journal publications about culture and human-computer interaction (HCI). In: *Human work interaction design: Usability in social, cultural and organizational contexts*. Berlin: Springer, 2010, pp. 98–112. DOI: http://dx.doi.org/10.1007/978-3-642-11762-6_9.

COHN, Neil. A visual lexicon. In: BOUISSAC, Paul, JORNA, René, NÖTH, Winfried, eds. *The Public Journal of Semiotics*. 1, 2007, pp. 35–56. ISSN 1918–9907. Available from: http://www.emaki.net/essays/visuallexicon.pdf.

COHN, Neil. ¡ *Eye graeflk Semiosis!* Thesis. The University of Chicago, 2005. Available from: `http://www.emaki.net/essays/visualsigns.pdf`.

COOPER, Alan, Robert REIMANN and David CRONIN. *About Face 3: The essentials of interaction design.* Indianapolis: Wiley Publishing, 2007, xxxv, 610 pp. ISBN 0470084111.

DE HAAN, Geert. *ETAG, a formal model of competence knowledge for user interface design.* 2000. Dissertation. Available from: `http://dspace.ubvu.vu.nl`.

DERRIDA, Jacques. *Texty k dekonstrukci: práce z let 1967–72.* Bratislava: Archa, 1993, 336 pp. ISBN 8071150460.

DE SOUZA, Clarisse Sieckenius, et al. The semiotic inspection method. In: *Proceedings of VII Brazilian symposium on Human factors in computing systems.* New York: ACM, 2006, pp. 148–57. ISBN:1595934324. DOI: `http://dx.doi.org/10.1145/1298023.1298044`.

DE SOUZA, Clarisse Sieckenius. *The semiotic engineering of human-computer interaction.* Cambridge, MA: MIT, 2005. ISBN 0262042207.

DILGER, Bradley. The ideology of ease. In: *Journal of Electronic Publishing.* 6 (September 2000). [cit. 2009-04-03]. ISSN 1080-2711. Available from: `http://quod.lib.umich.edu/cgi/t/text/text-idx?c=jep;view=text;rgn=main;idno=3336451.0006.104`.

DONG, Yin and LEE, Kun-Pyo. A cross-cultural comparative study of users' perceptions of a webpage: With a focus on the cognitive styles of Chinese, Koreans and Americans. *International Journal of Design* (2008) vol. 2 (2) pp. 19–30. Available from: `http://www.ijdesign.org/ojs/index.php/IJDesign/article/view/267/163`.

DRYER, Matthew S. Order of subject, object and verb. In: HASPELMATH, Martin, DRYER, Matthew S., GIL, David, COMRIE, Bernard, eds. *The world atlas of language structures online.* 2008. [cit. 2009-10-17.] Munich: Max Planck Digital Library, chapter 81. Available from: `http://wals.info/feature/81`.

ECO, Umberto. *Theory of semiotics.* Bloomington, IN: Indiana University Press, 1979, 354 pp. ISBN 9780253202178.

EHSES, Hanno. *Semiotic foundation of typography.* Halifax, Design Papers: Design Division, Nova Scotia College of Art and Design, 1976.

EISENSTEIN, Sergei. *Film Form: Essays in Film Theory.* Ed. and trans. Jay Leyda. Harcourt, Brace, 1969, 279 pp. ISBN [tbd].

EISENSTEIN, Sergei. *The Film Sense.* Ed. and trans. Jay Leyda. New York: Harvest, 1947, 1975, 288 pp. ISBN 0156309351.

ENGELHARDT, Jörg. *The language of graphics: A framework for the analysis of syntax and meaning in maps, charts and diagrams.* Amsterdam: Yuri Engelhardt, 2002, 197 pp. ISBN 9789057760891.

ENGELHARDT, Yuri. Syntactic structures in graphics. Computational Visualistics and Picture Morphology. In: SACHS-HOMBACH, Klaus, SCHIRRA, Jörg R. J., SCHWAN, Stephan, WULFF, Hans Jürgen, eds. *IMAGE—Zeitschrift för interdisziplinäre Bildforschung*, (5). 2007, 23–35. Verlag Herbert von Halem Verlagsgesellschaft, Köln. Available from: `http://www.gib.uni-tuebingen.de/own/journal/pdf/buch_image5b.pdf`, ISSN 1614-0885, Available from: `http://www.image-online.info/`.

ERICKSON, Thomas. Lingua Francas for design: Sacred places and pattern languages. In: *Proceedings of the 3rd Conference on Designing Interactive Systems: Processes, Practices, Methods, and Techniques.* New York: ACM, 2000, pp. 357–68. DOI: `http://dx.doi.org/10.1145/347642.347794`.

EVERS, Vanessa. Cross-cultural understanding of metaphors in interface design. In: ESS, C. and SUDWEEKS, F., eds. *Proceedings CATAC.* 1998, 98 pp. Available from: `https://lib.njnu.edu.cn/proxy/Enlinkwkkoxe001kxmyy11x 0vzqx0z1jixm1qs/~evers/pubs/catac98.pdf`.

Facebook [online]. Menlo Park, California, ©2004–. Available from: `http://facebook.com`.

FALLMAN, D. Persuade into what? Why Human-computer interaction needs a philosophy of technology. In: KORT, Y. de et al., ed. *PERSUASIVE 2007*, LNCS 4744, pp. 295–306. Berlin Heidelberg: Springer-Verlag, 2007. ISBN 9783540770053. DOI: `http://dx.doi.org/10.1007/978-3-540-77006-0_35`.

FLORIDI, Luciano. *Information: A very short introduction*. Oxford: Oxford University Press, 2010. ISBN 9780199551378.

FLUSSER, Vilém. *Komunikológia*. Bratislava: Mediálny inštitut, 2002, 253 pp. ISBN 8096877003.

FLUSSER, Vilém. *Do universa technických obrazů*. 1. vyd. Praha: Občanské sdružení pro podporu výtvarného umění, 2001, 162 pp. ISBN 8023875698.

FLUSSER, Vilém. *The shape of things: A philosophy of design*. London: Reaktion Books, 1999, 126 pp. ISBN 1861890559.

FLUSSER, Vilém and BOLLMANN, Stephan. *Lob der Oberflächlichkeit: für eine Phänomenologie der Medien*. Mannheim [u.a.], Bollmann, 1995, 336 pp. ISBN 9783927901360. (tran. KRTILOVÁ, Kateřina. Ineditum)

FOGG, Brian J. *Persuasive technology: Using computers to change what we think and do*. Amsterdam: Morgan Kaufmann Publishers, 2003, xxviii, 283 pp. ISBN 1558606432.

FOLEY, James D. and WALLACE, Victor L. The art of natural graphic man–machine conversation. In: *Proceedings of the IEEE*, 1974, 62.4: 462–71. DOI: `http://dx.doi.org/10.1109/PROC.1974.9450`.

FRASCA, Gonzalo. Videogames of the oppressed: Critical thinking, education, tolerance, and other trivial issues. In: WARDRIP-FRUIN, Noah, HARRIGAN, Pat., eds. *First-person: New media as story, performance and game*. Cambridge, MA: MIT, 2004, pp. 85–94. ISBN 0262731754.

GALLOWAY, Alexander R. The unworkable interface. *New literary history*, 2008, 39.4: 931–55. ISSN 0028-6087. DOI: `http://dx.doi.org/10.1353/nlh.0.0062`.

GARRETT, Jesse James. *Elements of user experience*. Indianapolis: New Riders, 2002, xiv, 189 pp. ISBN 0735712026.

GAVER, William W. Auditory icons: Using sound in computer interfaces. *Human–Computer Interaction*, 1986, 2.2: 167–77. DOI: `http://dx.doi.org/10.1207/s15327051hci0202_3`.

GE, Yan, et al. Chinese color preference in software design. In: *Engineering psychology and cognitive ergonomics*. Berlin: Springer, 2007, pp. 62–68. DOI: `http://dx.doi.org/10.1007/978-3-540-73331-7_7`.

GIBSON, James J. The concept of affordances. In: SHAW, Robert, BRANSFORD, John, eds. *Perceiving, acting, and knowing*. Hillsdale, NJ: Lawrence Erlbaum Associates; New York: distributed by the Halsted Press Division, Wiley, 1977, pp. 67–82. ISBN 0470990147.

GNU PROJECT. *GIMP—Development*. 2009. [cit. 2009-09-19]. Available from: `http://gimp.org/develop/`.

GNU PROJECT. *GIMP—The GNU Image Manipulation Program* [software]. Version 2.6.7. [Release date 13 August 2009.] Available from: `http://www.gimp.org/downloads/`.

GOOGLE. *Google Earth* [software]. Version 5.2.1.1588. [Release date 03 September 2010.] Mountain View, CA. Available from: `http://www.google.com/earth/`.

GRICE, H. Paul. Logic and conversation. In: COLE, Peter and MORGAN, Jerry L., eds. *Syntax and Semantics*, Vol. 3, Speech Acts. New York: Academic Press, 1975, pp. 41–58 [here 45–47]. ISBN 0127854231.

GUEST, Ann Hutchinson. *Labanotation: The system of analyzing and recording movement.* 4th ed., rev. New York: Routledge, 2005, xiv, 487 pp. ISBN 0415965624.

GUEST, Greg, BUNCE, Arwen and JOHNSON, Laura. How many interviews are enough? An experiment with data saturation and variability. *Field methods*, 2006, 18.1: 59–82. DOI: http://dx.doi.org/10.1177/1525822X05279903.

HALL, Stuart. Signification, representation, ideology: Althusser and the post-structuralist debates. *Critical Studies in Media Communication*, 1985, 2.2: 91–114. DOI: http://dx.doi.org/10.1080/15295038509360070.

HALLIDAY, Michael Alexander Kirkwood. *An introduction to functional grammar.* Reprinted. London: Arnold, 1985, xxxv, 387 pp. ISBN 0713163658.

HÉBERT, Louis. The Actantial model. *Signo: Theoretical Semiotics on the Web* [online]. Rimouski (Quebec), 2006. [cit. 2009-09-24]. Available from: http://www.signosemio.com/greimas/actantial-model.asp.

HJELMSLEV, Louis. *Prolegomena to a theory of language.* Madison: University of Wisconsin Press, 1961, 144 pp.

HOFSTEDE, Geert, HOFSTEDE, Jan and MINKOV, Michael. Cultures and organizations: Software of the mind; intercultural cooperation and its importance for survival. 3rd ed. New York: McGraw-Hill, 2010, 561 pp. ISBN 9780071664189.

HORN, Robert E. *Visual language: Global communication for the 21st century.* MacroVU, Inc., 1998, 270 pp. ISBN 9781892637093.

HOTCHKISS, Gord. Chinese eye tracking study: Baidu vs Google. In: *Search Engine Land* [online]. 15 Jun 2007. [cit. 2011-07-13]. Available from: http://searchengineland.com/chinese-eye-tracking-study-baidu-vs-google-11477.

JACOBSON, Ivar et al. *Object-oriented software engineering: A use case driven approach.* New York: ACM Press; Wokingham, UK/Reading, MA: Addison-Wesley Pub., 1992, 524 pp. ISBN 0201544350.

Kayak [online]. Norwalk, Connecticut, ©2013 KAYAK.com. Available from: http://www.kayak.com.

KRESS, Gunther and VAN LEEUWEN, Theo. *Reading images: The grammar of visual design.* 2nd ed. London: Routledge, 2006, xv, 291 pp. ISBN 0415319153.

KURNIAWAN, Sri H., GOONETILLEKE, Ravindra S. and SHIH, Heloisa M. Involving Chinese users in analyzing the effects of languages and modalities on computer icons. In: STEPHANIDIS, Constantine. (2001), ed. *Universal access in HCI: Towards an information society for all: Volume 3 of the proceedings of HCI International 2001*, pp. 491–494. Mahwah: Lawrence Erlbaum, 2001, 1133 pp. ISBN 0805836098.

LANHAM, Richard A. *A handlist of rhetorical terms.* 2nd ed. Berkeley: University of California Press, 1991, xv, 205 pp. ISBN 9780520076693.

LAUREL, Brenda. *Computers as theatre.* Reading, MA: Addison-Wesley, 1993, xxv, 227 pp. ISBN 0201550601.

LEVY, Steven. *Hackers: Heroes of the computer revolution.* Garden City, NY: Anchor Press/Doubleday, 1984, xv, 458 pp. ISBN 0385191952.

LISZKA, James Jakób. *A general introduction to the semeiotic of Charles Sanders Peirce.* Bloomington, IN: Indiana University Press, 1996, 151 pp. ISBN 9780253330475.

LOKE, Lian, LARSSEN, Astrid T. and ROBERTSON, Toni. Labanotation for design of movement-based interaction. In: *Proceedings of the Second Australasian Conference on Interactive Entertainment.* Creativity & Cognition Studios Press, 2005. pp. 113–20. ISBN 0975153323.

LUHMANN, Niklas. *Social systems.* Stanford: Stanford University Press, 1995, x, 627 pp. ISBN 0804726256.

MARCUS, Aaron. The money machine: Helping baby boomers retire. In: *User Experience*. 2012a, vol. 11 (2) pp. 24–27. Available from: `http://www.UsabilityProfessionals.org`.

MARCUS, Aaron. The story machine: A mobile app to change family story-sharing behavior. In: *Proceedings of CHI 2012*. 2012b, pp. 1–4.

MARCUS, Aaron. The health machine: Mobile UX design that combines information design with persuasion design. In: *Design, user experience, and usability: Theory, methods, tools and practice*. Berlin: Springer, 2011. pp. 598–607. DOI: `http://dx.doi.org/10.1007/978-3-642-21708-1_67`.

MARCUS, Aaron. Integrated information systems: A professional field for information designers. In: *Information Design Journal*, 2009, 17.1: 4–21. DOI: `http://dx.doi.org/10.1075/idj.17.1.02mar`.

MARCUS, Aaron. *UI center of excellence handbook*. Berkeley: Aaron Marcus and Associates, Inc., 2003a. Internal document.

MARCUS, Aaron. Icons, symbols, and signs: Visible languages to facilitate communication. In: *Interactions*, 2003b, 10.3: 37–43. DOI: `http://dx.doi.org/10.1145/769759.769774`.

MARCUS, Aaron. Dare we define user-interface design? *interactions*, 2002, 9.5: 19–24. DOI: `http://dx.doi.org/10.1145/566981.566992`.

MARCUS, Aaron. Cross-cultural user-interface design. In: SMITH, Michael J., and SALVENDY, Gavriel, eds. *Proceedings, Vol. 2, Human–Computer Interface Internat. (HCII) Conf.*, 5–10 Aug., 2001, New Orleans, LA, USA. Mahwah, NJ: Lawrence Erlbaum Associates, 2001, pp. 502–505. vol. 2. ISBN 1581135378. DOI: `http://dx.doi.org/10.1145/1556262.1556264`.

MARCUS, Aaron. Metaphor design for user interfaces. In: *CHI 98 Conference Summary on Human Factors in Computing Systems*. New York: ACM, 1998, pp. 129–130. ISBN 1581130287. DOI: `http://dx.doi.org/10.1145/286498.286577`.

MARCUS, Aaron. *Graphic design for electronic documents and user interfaces*. New York: ACM Press; Reading, MA: Addison-Wesley, 1992, 266 pp. ISBN 0201543648.

MARCUS, Aaron. Visual rhetoric in a pictographic-ideographic narrative. In: BORBÉ, Tasso. *Semiotics unfolding: Proceedings of the Second Congress of the International Association for Semiotic Studies, Vienna, July 1979 Vol. III*. Semiotics Unfolding. Berlin: Mouton, 1983, pp. 1500–8. ISBN 9783110097795.

MARCUS, Aaron and GOULD, Emilie West. Crosscurrents: Cultural dimensions and global Web user-interface design. In: *interactions*, 2000, 7.4: 32–46. DOI: `http://dx.doi.org/10.1145/345190.345238`.

MARCUS, Aaron and JEAN, Jérémie. Going green at home: The green machine. In: *Information Design Journal*, 2009, 17.3: 235–45. DOI: `http://dx.doi.org/10.1075/idj.17.3.08mar`.

MASON, Mark. Sample size and saturation in PhD studies using qualitative interviews. In: *Forum Qualitative Sozialforschung/Forum: Qualitative Social Research*. 2010. ISSN 1438-5627. Available from: `http://www.qualitative-research.net/index.php/fqs/article/download/1428/3028`.

MCCLOUD, Scott. *Understanding comics: The invisible art*. New York: HarperPerennial, 1994, 215 pp. ISBN 006097625x.

MERRIAM-WEBSTER. *Data—Definition and more from the free Merriam-Webster Dictionary* [online]. ©2013 Merriam-Webster, Incorporated. [cit. 2013-01-19]. 2013a. Available from: `http://www.merriam-webster.com/dictionary/data`.

MERRIAM-WEBSTER. *Design—Definition and more from the free Merriam-Webster Dictionary* [online]. ©2013 Merriam-Webster, Incorporated. [cit. 2013-01-19]. 2013b. Available from: `http://www.merriam-webster.com/dictionary/design`.

MERRIAM-WEBSTER. *Inform—Definition and more from the free Merriam-Webster Dictionary* [online]. ©2013 Merriam-Webster, Incorporated. [cit. 2013-01-19]. 2013c. Available from: `http://www.merriam-webster.com/dictionary/inform`.

MERRIAM-WEBSTER. *Information science—Definition and more from the free Merriam-Webster Dictionary* [online]. ©2013 Merriam-Webster, Incorporated. [cit. 2013-01-19]. 2013d. Available from: `http://www.merriam-webster.com/dictionary/information\%20science`.

MERRIAM-WEBSTER. *Information theory—Definition and more from the free Merriam-Webster Dictionary* [online]. ©2013 Merriam-Webster, Incorporated. [cit. 2013- 01-19]. 2013e. Available from: `http://www.merriam-webster.com/dictionary/information\%20theory`.

MERRIAM-WEBSTER. *Logos—Definition and more from the free Merriam-Webster Dictionary* [online]. ©2013 Merriam-Webster, Incorporated. [cit. 2013-01-19]. 2013f. Available from: `http://www.merriam-webster.com/dictionary/logos`.

METZ, Christian. *Film language: A semiotics of the cinema*, trans. Michael Taylor. New York: Oxford University Press, 1974, xiv, 268 pp.

MICROSOFT CORPORATION. *Microsoft Windows XP SP3*. Redmond, California. 2008.

MILLER, George. The magical number seven, plus or minus two: Some limits on our capacity for processing information. *The Psychological Review*, 1956, 63: 81–97. Available from: `http://cogprints.org/730/1/miller.html`.

MITCHELL, W. J. Thomas. *Iconology: Image, text, ideology*. Chicago: University of Chicago Press, 1986, x, 226 pp. ISBN 9780226532295.

MOLES, Abraham A. *Information theory and esthetic perception*. Urbana: University of Illinois Press, 1966. 217 pp.

MORRIS, C. William. *Foundations of the theory of signs*. Chicago: University of Chicago Press, 1970. ISBN 0226532321.

MULLINS, Willard A. On the concept of ideology in political science. In: *The American Political Science Review* 66 (1972): 498–510. ISSN 0003-0554. DOI: `http://dx.doi.org/10.2307/1957794`.

NADIN, Mihai. Interface design: A semiotic paradigm. In: *Semiotica*, 1988, 69.3–4: 269–302. ISSN (Online) 1613-3692, ISSN (Print) 0037-1998, DOI: `http://dx.doi.org/10.1515/semi.1988.69.3-4.269`, November 2009.

NARAYANAN, N. Hari and HÜBSCHER, Roland. Visual language theory: Towards a human-computer interaction perspective. In: *Visual language theory*. New York: Springer, 1998, pp. 87–128. DOI: `http://dx.doi.org/10.1007/978-1-4612-1676-6_3`.

NIELSEN, Jakob. Mastery, mystery, and misery: The ideologies of web design. In: *UseIt.com* [online]. 2004-08-30. [cit. 2009-04-03]. ISSN 1548-5552. Available from: `http://www.useit.com/alertbox/20040830.html`.

NIELSEN, Jakob. Usability inspection methods. In: *Conference companion on Human factors in computing systems*. New York: ACM, 1994, pp. 413–414. ISBN 0897916514. DOI: `http://dx.doi.org/10.1145/259963.260531`.

NIELSEN, Jakob and LANDAUER, Thomas K. A mathematical model of the finding of usability problems. In: *Proceedings of the INTERACT'93 and CHI'93 conference on Human factors in computing systems*. New York: ACM, 1993, pp. 206–213. ISBN 0897915755. DOI: `http://dx.doi.org/10.1145/169059.169166`.

NIELSEN NORMAN GROUP. *Heuristic evaluation articles and training* [online]. [cit. 2013-06-02]. Available from: `http://www.nngroup.com/topic/heuristic-evaluation`.

NISBETT, Richard. *The geography of thought: How Asians and Westerners think differently... and why*. New York: Free Press, 2004, xxiii, 264 pp. ISBN: 0743255356.

NISBETT, Richard E. and MASUDA, Takahiko. Culture and point of view. *Proceedings of the National Academy of Sciences*, 2003, 100.19: 11163–11170. DOI: http://dx.doi.org/10.1073/pnas.1934527100.

NORMAN, Donald A. *The design of everyday things*. Basic Books (AZ), ©1988. New York: Basic Books, 2002, xxi, 257 pp. ISBN: 0465067107.

NORMAN, Donald A. Affordance, conventions, and design. *interactions*, 1999, 6.3: 38–43. DOI: http://dx.doi.org/10.1145/301153.301168.

NORMAN, Donald A. Cognitive engineering. In: NORMAN, D. A. and DRAPER, S. W., eds. *User centered system design: New perspectives on human–computer interaction*. Hillsdale, NJ: Lawrence Erlbaum and Associates, 1986, pp. 31–61, xiii, 526 pp. ISBN: 0898597811.

NÖTH, Winfried. *Handbook of semiotics*. Bloomington, IN: Indiana University Press, 1995, xii, 576 pp. ISBN: 0253209595.

O'NEILL, Shaleph, BENYON, D., and TURNER, Susan. Semiotics and interaction analysis. In: *Proceeding of the 11th European Conference on Cognitive Ergonomics (ECCE-11)*, Catania, Sicily: University of Teesside, 2002.

Open Source Initiative [online]. [cit. 2009-06-11]. Available from: http://www.opensource.org.

PAYNE, Stephen J. Interface problems and Interface resources. In: CARROLL, John. M., ed. *Designing interaction: Psychology at the human-computer interface*. Cambridge, UK: Cambridge University Press, 1991, p. 136, ix, 333 pp. ISBN: 0521400562.

PAYNE, Stephen J. and GREEN, Thomas R.G. Task-action grammars: A model of the mental representation of task languages. In: *Human–computer interaction*, 1986, 2.2: 93–133. DOI: http://dx.doi.org/10.1207/s15327051hci0202_1.

PEARSON, Charls R. and SLAMECKA, Vladimir. *Semiotic foundations of information science*. Georgia Institute of Technology, School of Information and Computer Science, 1977. Available from: http://smartech.gatech.edu/jspui/bitstream/1853/40594/1/g-36-611_119578.pdf.

PIERCE, John R. *An introduction to information theory: Symbols, signals and noise*. Mineola, NY: Courier Dover Publications, 1980.

PIMENTA, Marcelo Soares and FAUST, Richard. Eliciting interactive systems requirements in a language-centered user-designer collaboration. *ACM SIGCHI Bulletin*, 1997, 29.1: 61–65. Available from: http://old.sigchi.org/bulletin/1997.1/pimenta.html.

RABER, Douglas and BUDD, John M. Information as sign: Semiotics and information science. In: *Journal of Documentation*, 2003, 59.5: 507–22. ISSN: 0022-0418. DOI: http://dx.doi.org/10.1108/00220410310499564.

RAJ, Avinash and KOMARAGIRI, Vihari. RUCID: Rapid usable consistent interaction design patterns-based mobile phone UI design library, process and tool. In: *Human–computer interaction: New trends*. Berlin: Springer, 2009. pp. 677–686. DOI: http://dx.doi.org/10.1007/978-3-642-02574-7_76.

RHEINFRANK, John and EVENSON, Shelley. Design languages. In: WINOGRAD, Terry. *Bringing design to Software*. New York: ACM, 1996, pp. 63–85. ISBN 0201854910.

RICHARDS, Clive J. Diagrammatics: An investigation aimed at providing a theoretical framework for studying diagrams and for establishing a taxonomy of their fundamental modes of graphic organization. Ph.D. thesis, Royal College of Art, London. 1984.

ROAM, Dan. *Blah, blah, blah: What to do when words don't work*. New York: Portfolio/Penguin, 2011, xi, 350 pp. ISBN: 1591844592.

SAPIR, Edward and MANDELBAUM, David Goodman. *Selected writings of Edward Sapir in language, culture and personality*. Berkeley: University of California Press, 1949, xv, 617 pp. ISBN: 0520055942.

SAURO, Jeff. Confidence interval calculator for a completion rate. In: *Measuring Usability* [online]. 1 Oct 2005 [cit. 2012-09-01]. Available from: `http://www.measuringusability.com/wald.htm`.

SCHÜTZ, Alfred and LUCKMANN, Thomas. *Structures of the life world, Vol 1.* Evanston, IL: Northwestern University Press, 1973–c1989. ISBN 0810106221.

SEARLE, John R. *Making the social world: The structure of human civilization.* New York: Oxford University Press, 2010, xiv, 208 pp. ISBN 9780195396171.

SEARLE, John R. *Consciousness and language.* Cambridge, UK: Cambridge University Press, 2002, vii, 269 pp. ISBN 0521597447.

SEARLE, John R. *The construction of social reality.* New York: Free Press, 1995, xiii, 241 pp. ISBN: 9780684831794.

SEARLE, John R. *Speech acts: An essay in the philosophy of language.* Cambridge, UK: Cambridge University Press, 1969, vii, 203 pp. ISBN 052109626X.

SENGERS, Phoebe. The Ideology of Modernism in HCI. ACM CHI 2010 Workshop on Critical Dialogue: Interaction, Experience, Theory, Atlanta, GA, April 2010. Also avaiable from: `http://www.cl.cam.ac.uk/events/experiencingcriticaltheory/Sengers-IdeologyModernism.pdf`.

SHANNON, Claude E. A Mathematical theory of communication. In: *Bell System Technical Journal 27. 1948* (July and October): pp. 379–423, 623–56.

SHARP, Helen, ROGERS, Yvonne and PREECE, Jenny. *Interaction design: Beyond human–computer interaction*, 2nd edition. Chichester: John Wiley and Sons, 2007. ISBN: 9780470018668.

SHEN, Siu-Tsen, WOOLLEY, Martin and PRIOR, Stephen. Towards culture-centred design. In: *Interacting with computers*, 2006, 18.4: 820–852. DOI: `http://dx.doi.org/10.1016/j.intcom.2005.11.014`.

SHERIDAN, E. F. Cross-cultural web site design: Considerations for developing and strategies for validating locale appropriate on-line content. In: *MultiLingual Computing & Technology*, 2001, 43.12: 7, pp. 1–5. Available at: `http://www.multilingual.com/articleDetail.php?id=595`.

SMITH, Andy, et al. A process model for developing usable cross-cultural websites. In: *Interacting with computers*, 2004, 16.1: 63–91. DOI: `http://dx.doi.org/10.1016/j.intcom.2003.11.005`.

SMITH, Andy. Issues in adapting usability testing for global usability. In: *Global Usability*. London: Springer, 2011, pp. 23–38. DOI: `http://dx.doi.org/10.1007/978-0-85729-304-6_3`.

SurveyGizmo [online]. Widgix, LLC dba SurveyGizmo, ©2005–2013. Available from: `http://www.surveygizmo.com`. Online survey software.

SOMOL, Robert and WHITING, Sarah. Notes around the Doppler Effect and other Moods of Modernism. *Perspecta*, 2002, 33: 72–77. Published by: The MIT Press on behalf of Perspecta. Available from: `http://www.jstor.org/stable/1567298`.

SUTHERLAND, William Robert. *On-line graphical specification of computer procedures.* Dissertation. Lincoln Lab Mass. Inst. of Tech. Lexington, 1966. Available from: `http://dspace.mit.edu/bitstream/handle/1721.1/13474/25697177.pdf`.

SUTNAR, Ladislav. *Visual design in action: Principles, purposes.* New Yrok: Hastings House, 1961.

TAVASSOLI, Nader T. *Beyond reading: Visual processing of language in Chinese and English.* MIT Working Paper, 2002. Available from: `http://www.bus.umich.edu/facultyresearch/researchcenters/centers/yaffe/downloads/Complete_List_of_Working_Papers/nader.pdf`.

TIDWELL, Jenifer. *Designing interfaces.* Sebastopol: O'Reilly, 2006, xx, 331 pp. ISBN 0596008031.

TOGNAZZINI, Bruce. *TOG on interface*. Reading, MA: Addison-Wesley, 1992, xvi, 331 pp. ISBN: 9780201608427.

TRACY, De, Antoine Louis Claude Destutt. [Edited by JEFFERSON, Thomas]. A treatise on political economy. [1817]. Auburn, AL: The Ludwig von Mises Institute, 2009.

TUFTE, Edward R. *Visual explanation: Images and quantities, evidence and narrative.* Cheshire, CT: Graphics Press, 1997. ISBN 0961392126.

TUFTE, Edward R. *Envisioning information.* Cheshire: Graphics Press, 1991, 126 pp. ISBN 9780961392116.

TUFTE, Edward R. *The visual display of quantitative information.* Cheshire: Graphics Press, 1983, 197 pp. ISBN 096139210X.

VAN DUYNE, Douglas K., LANDAY, James A. and HONG, Jason I. *The design of sites: Patterns, principles, and processes for crafting a customer-centered Web experience*, 2nd edition. Reading, MA: Addison-Wesley Professional, 2006.

WAI, Conrad and MORTENSEN, Pete. Persuasive technologies should be boring. In: *Persuasive technology*. Berlin: Springer, 2007, pp. 96–99. ISBN 9783540770053. DOI: `http://dx.doi.org/10.1007/978-3-540-77006-0_12`.

WHORF, Benjamin Lee. CARROLL, John B, LEVINSON Stephen C., LEE, Penny, eds. *Language, thought, and reality: Selected writings of Benjamin Lee Whorf.* 2nd edition. Cambridge, MA: MIT Press, 2012, xxiii, 417 pp. ISBN 9780262517751.

Wikipedia: The free encyclopedia [online]. San Francisco, California: Wikimedia Foundation, ©2001–[cit. 2009-06-11]. English version. Available from: `http://en.wikipedia.org/`.

WINOGRAD, Terry. A language/action perspective on the design of cooperative work. In: *Human-Computer Interaction 3:1* (1987–88), 3–30. DOI: `http://dx.doi.org/10.1207/s15327051hci0301_2`.

WINOGRAD, Terry and FLORES, Fernando. *Understanding computers and cognition: A new foundation for design.* Reading, MA: Addison-Wesley Publishing Company, 1987, 207 pp. ISBN 9780201112979.

WITTGENSTEIN, Ludwig. *Philosophical investigations.* 3rd edition. Oxford: Blackwell, 1986 c1958, viii, 250 pp. ISBN 0631146709.

WITTGENSTEIN, Ludwig. *Tractatus logico-philosophicus.* London: Kegan Paul, Trench, Trubner & Co., Ltd.; New York: Harcourt, Brace & Company, Inc., 1922. Project Gutenberg's Tractatus Logico-Philosophicus, eBook. Available from: `http://www.gutenberg.org/ebooks/5740`.

WORLD WIDE WEB CONSORTIUM (W3C). *Basic HTML data types* [online]. [cit. 2014-06-18]. Available from `http://www.w3.org/TR/REC-html40/types.html\#h-6.5`.

Index